Islamic Republic of Iran: Managing the Transition to a Market Economy

Abdelali Jbili, Vitali Kramarenko, and José Bailén

International Monetary Fund
Washington, DC

Production: IMF Multimedia Services Division
Cover Design: Massoud Etemadi
Typesetting: Alicia Etchebarne-Bourdin
Cover Photo: Getty Images

Cataloging-in-Publication Data

Jbili, A. (Abdelali).
Islamic Republic of Iran: managing the transition to a market economy/Abdelali Jbili, Vitali Kramarenko, and José Bailén—[Washington, D.C.: International Monetary Fund, 2007]

p. cm

Includes bibliographical references.
ISBN 978-1-58906-441-6

1. Iran—Economic conditions. 2. Monetary policy—Iran. 3. Foreign exchange rates—Iran. 4. Petroleum industry and trade—Iran. I. Kramarenko, V. (Vitali). II. Bailén, J.M. (José María). III. International Monetary Fund.
HC473.J25 2007

Price: $27.00

Please send orders to:
International Monetary Fund, Publication Services
700 19th Street, NW, Washington, DC 20431, U.S.A.
Telephone: (202) 623-7430 Telefax: (202) 623-7201
Internet: http://www.imf.org

Contents

Appendixes

Boxes

Tables

Figures

Acknowledgments

The authors are grateful for the useful comments and suggestions made by Mohammad Jafar Mojarrad and Akbar Komijani, Vice Governors of the Central Bank of Iran (CBI); Abbas Mirakhor, IMF Executive Director for Iran; and by senior staff of the CBI and Management and Planning Organization. These comments and suggestions were made during a seminar presentation of these chapters at the CBI in June 2004. The authors would like to thank Lorenzo Pérez, Edward Gardner, and Hamid Davoodi for insightful comments. Abhisek Banerjee provided excellent research assistance. Audrey Scott and Fiorella Facello compiled the document and provided invaluable administrative assistance. Special thanks go to Linda Griffin Kean and Alicia Etchebarne-Bourdin in the External Relations Department of the Fund for their contributions in editing and coordinating the publication of the paper.

Abbreviations

ADF	Augmented Dickey-Fuller
AML	anti–money laundering
CBI	Central Bank of Iran
CBPP	central bank participation paper
CPI	consumer price index
FDI	foreign direct investment
FSAP	Financial Sector Assessment Program
FYDP	five-year development plan
GPP	government participation paper
HLS	half-life of a unit shock
IFS	International Financial Statistics (IMF)
LIBOR	London interbank offered rate
MCC	Money and Credit Council
MENA	Middle East and North Africa
NEER	nominal effective exchange rate
NFPE	nonfinancial public enterprise
OLS	ordinary least squares
OSF	Oil Stabilization Fund
P/E ratio	price-earnings ratio
PPP	purchasing power parity
REER	real effective exchange rate
Rls	rials
TFP	total factor productivity
TPID	Trade Policy Information Database (Policy Development and Review Department, IMF)
TSE	Tehran Stock Exchange
VAT	value-added tax
WEO	World Economic Outlook (IMF)

Preface

Most of the chapters in this paper are based on analytical work undertaken in preparation for the IMF's 2004 Article IV consultation with the Iranian authorities. Since then, a number of changes have taken place at the domestic level and with regard to the Islamic Republic of Iran's external economic environment. These changes do not materially alter the paper's findings and conclusions, but may have implications for future economic policymaking in Iran. Without undertaking a full-fledged updating of the paper, it is useful to highlight up front key developments and policy changes and their likely effects.

First, as a major oil producer, Iran has benefited from the sharp increase in oil prices in recent years. The average export price of Iranian crude almost doubled between 2002/03 and 2005/06, rising from about $27 per barrel to almost $51 per barrel. Iran's export earnings from oil and gas reached close to $50 billion in 2005/06. Such a large positive shock has further improved an already strong external position, as reflected in the growing current account surplus, the large buildup of international reserves, and the low external debt ratio. Assuming that part of the increase in oil revenue will be permanent, Iran's external resources, if adequately managed, are sufficient to meet its development objectives and build savings for future generations.

Second, since 2002/03, Iran has made major strides in opening its economy to the rest of the world through trade liberalization, unification of the exchange rate, elimination of exchange restrictions, and attraction of foreign direct investment (FDI). These efforts have culminated in Iran's acceptance of Article VIII of the IMF's Articles of Agreement,[1] its return to international financial markets from which it had been absent for almost three decades, and its application for membership in the World Trade Organization. There is no doubt that these policies contributed to the favorable performance highlighted in this paper and that Iran would benefit from further integration into the global economy.

Third, the Iranian economy has sustained high growth for the past five years, but reducing unemployment remains a key challenge. Against the background of abundant oil revenue, real GDP growth has remained strong and broad-based. Despite some deceleration in 2004/05 owing to adverse

[1]Under Article VIII, Sections 2 (a) and 3 of the IMF's Articles of Agreement, Iran committed to eliminating all exchange restrictions on current account transactions and refraining from introducing new ones.

weather conditions and capacity constraints in the oil sector, real GDP growth was 6¼ percent a year, on average, during the period 2002/03–2005/06. Unemployment has continued to trend down but remains high at about 10 percent.

Fourth, buoyed by high oil revenue, fiscal policy has been expansionary. The non-oil fiscal deficit remained close to 24.5 percent of non-oil GDP in 2005/06, roughly unchanged from 2002/03. This, in turn, has put upward pressure on domestic liquidity and inflation. The inflation rate, as measured by the consumer price index, averaged 14 percent a year during the five years up to 2005/06. The authorities have committed to reduce inflation to a single-digit level by 2010/11, the end of the current five-year development plan (FYDP), but this objective has so far proven elusive. Recent attempts to bring down inflation through price freezes are only a temporary remedy.

Finally, the overall setting for macroeconomic policy has undergone some changes since 2003/04. In the financial sector, new private banks have been licensed, bank supervision has been strengthened, and new capital market regulations have been introduced. Banks have been given more leeway in allocating credit, but most rates of return (equivalent to interest rates) have continued to be set administratively, and little progress has been made toward developing effective and market-based monetary policy instruments. Moreover, efforts to strengthen the central bank's autonomy have been weakened by parliament's increased oversight over the use of some monetary policy instruments.

These developments highlight the need for macroeconomic policies to better focus on helping contain the growth of domestic demand, reduce fiscal dominance, curb the rapid growth of bank credit, and contain inflationary pressures. The Iranian authorities cannot rely indefinitely on fiscal and monetary expansion to boost growth and employment. Instead, there is a need to direct resources away from wasteful subsidies toward high-quality projects in infrastructure and social services. This, together with reforms aimed at improving productivity and economic efficiency, would encourage the private sector to grow and foster a sustainable increase in employment.

The Fourth Five-Year Development Plan, introduced in 2004/05, aims to raise real GDP growth to about 8 percent a year on average and calls for far-reaching structural reforms, including privatization, financial sector reform, an overhaul of the large subsidies, and development of technology- and knowledge-based activities. Although preparations for some of these reforms (such as streamlining the subsidy system) are under way, the speed and depth of implementation need to be accelerated substantially to achieve the plan's objectives.

Introduction

Iran has received much attention from a geopolitical and regional standpoint, but its economic challenges have not attracted a similar degree of interest. Yet, with a population of 69 million, considerable hydrocarbon resources, a dynamic and entrepreneurial middle class, and a relatively well-educated labor force, Iran's economic potential is considerable. Equally important is Iran's experience with economic reform, which deserves attention for several reasons:

- Iran is in transition to a market economy, and the process has been neither smooth nor supported by external conditions. An assessment of economic performance since the 1979 revolution must be viewed against the historical background of several years of political and institutional upheaval, an eight-year war with neighboring Iraq (1980–88), regional instability, trade sanctions, and two major episodes of boom and bust in oil prices. These events have left their mark on the economy—not to mention the related human cost—through a sharp contraction in real GDP, destruction of physical infrastructure and property, high inflation, and food and supply shortages. From this perspective, the Iranian experience is one of resilience.

- Significant advances have been made in recent years in opening up the economy to international trade and FDI, encouraging private sector activities, removing exchange restrictions, reforming the tax system, and enhancing macroeconomic management. These reforms have started to bear fruit, as evidenced by the pickup in growth, the improvement in other key macroeconomic indicators, increased business confidence, and Iran's renewed access to international capital markets.

- Despite the progress, major challenges remain. The economy continues to be saddled with large and inefficient subsidies, and private sector development is hindered by administrative controls and rigidities. Unemployment is high, and inflation remains in double digits. The next phase of reforms would need to strengthen the foundations of private-sector-led growth in a context of greater integration with the world economy in order to achieve higher levels of efficiency and competitiveness. The reforms would also need to address fiscal policy challenges arising from the management of Iran's large oil and gas reserves, promote diversification of the economy, and improve macroeconomic policies to reduce inflation to single-digit levels over the medium term. These reforms are critical to help sustain high growth and employment creation.

- The Iranian experience offers a remarkable example of ownership of reforms. Despite years of economic isolation, Iran has remained open to advice and technical assistance from bilateral and multilateral sources, although the solutions to its specific economic problems have been largely homegrown. This openness to external advice, in turn, has increased the confidence of the countries' economic managers and brought about further opening to the global economy, which has led to substantial improvements in transparency and the dissemination of economic and financial statistics.

This IMF Special Issues Paper takes stock of key developments in the Iranian economy in recent years, reviews the major issues and policy responses, highlights the nature of the challenges faced by Iran, and draws implications for the next phase of reforms.

Chapter 1 takes a historical view of economic growth in Iran (1960–2002), analyzing the role of capital accumulation, investment in human capital, and total factor productivity (TFP) in explaining Iran's growth performance. It also attempts to quantify the role of several other factors—such as macroeconomic stability, financial sector development, trade openness, and the terms of trade—and discusses the impact of noneconomic factors, such as political developments. The analysis highlights the fact that the Iranian economy can sustain high growth rates for an extended period, and it also provides evidence of the negative effects on growth of inflation, widespread distortions, and inward-oriented economic policies that tend to lower TFP. The lessons to be drawn from this review are relevant for the period ahead, given the expectation that about one-third of the growth rate targeted by the Fourth Five-Year Development Plan (2005/06–2010/11) is to come from higher TFP.

Chapter 2 focuses on financial sector development, which is a key building block of a high-growth strategy. Starting with a comprehensive review of the current status of the financial system and its evolution, the chapter highlights the reasons behind the low level of financial sector development in Iran. It draws on the literature and cross-country experience to show the links between financial sector reform and the performance of the real economy. Finally, the chapter points to key areas in which reform is urgently needed, including regulation and supervision, bank restructuring and management, and the development of capital markets.

This analysis sets the stage for Chapter 3, which contains an in-depth discussion of monetary policy in light of the gradual move in recent years toward market-based monetary and exchange rate policies. The discussion provides helpful insights into the complexity of the Iranian monetary policy framework, which is based on Islamic finance principles and is characterized by, among other factors, fiscal dominance, lack of market-based instruments of liquidity management, and limitations on the independence of the central bank in pursuing its monetary policy objectives.

Chapter 4 provides a brief review of the institutional setup of the foreign exchange market during 1993–2003 and examines a range of indicators that could be used to assess the exchange rate level. After a long period of multiple exchange rates, Iran unified the exchange rate and established a managed float exchange rate regime in 2002. Finding appropriate indicators to assess the exchange rate level, let alone to estimate an equilibrium real effective exchange rate (REER), presents a major challenge in the Iranian context. The difficulties stem from the large share of oil exports in trade, various distortions arising from government intervention in economic activity, and the relatively short experience with a unified market-based exchange rate. Standard approaches to assessing the exchange rate level based on the purchasing power parity (PPP) theory or on an analysis of productivity differentials and terms of trade show inconclusive results. Thus, this chapter relies on a number of indicators of performance of the tradables sector, as well as an assessment of equilibrium external balances, to provide additional insights into formulating a judgment on the current level of the exchange rate.

Chapter 5 discusses fiscal management in an oil economy, with an application to Iran, and highlights various options for developing a medium-term fiscal framework. The chapter outlines the long-term objective of preserving hydrocarbon wealth for the benefit of future generations and discusses other short-term objectives of fiscal policy, including the need to bring down inflation and reduce the economy's vulnerability to a decline in oil prices. On the basis of the prices prevailing in 2003/04 and current policies, the analysis indicates that Iran is overconsuming its oil wealth resources and that a comprehensive energy price reform could play a critical role in improving welfare and intergenerational equity.

CHAPTER

1 Economic Growth

Iran faces the challenge of increasing its growth rate to reduce unemployment and improve the living standards of its population over the medium term. Growth performance in recent years (6 percent during 2000–03) has been satisfactory and has been driven by major economic reforms as well as transitory factors such as high oil prices and expansionary fiscal and monetary policies. However, questions arise about the determinants of growth in Iran and the long-term sustainability of relatively high growth rates. Because past experience shows that the Iranian economy can grow at relatively high rates over an extended period, the first step is to examine the historical sources of growth and discuss the relevance of various contributing factors for the medium term. The second step is to provide an analytical framework for the formulation of growth-enhancing policies.

This chapter uses a growth accounting exercise to quantify the historical sources of growth during the period 1960–2002, including human capital accumulation and the contribution of TFP to growth. The chapter also presents an empirical study to quantify the role of several other contributing factors commonly discussed in the cross-country growth literature, including macroeconomic stability, financial development, trade openness, and changes in the terms of trade.[1]

Historical View of Growth Performance

During 1960–2002, real GDP growth in Iran averaged 4.6 percent per year (2 percent in per capita terms). Non-oil GDP grew at a faster pace of 5.5 percent during this period.[2] There are three distinct subperiods (Figure 1):

Note: Prepared by José Bailén.

[1]Recent growth studies on Iran include Jalali-Naini (2003) and Mojaverhosseini (2003).

[2]Since the relative price of oil GDP increased by an average of 3 percent per year during the period 1960–2002, the ratio of nominal oil GDP to total GDP increased from 12.8 percent to 22.1 percent, even though real oil output increased at a slower pace than real non-oil output.

Figure 1. GDP Growth Rates, 1960–2002

(Annual percentage change)

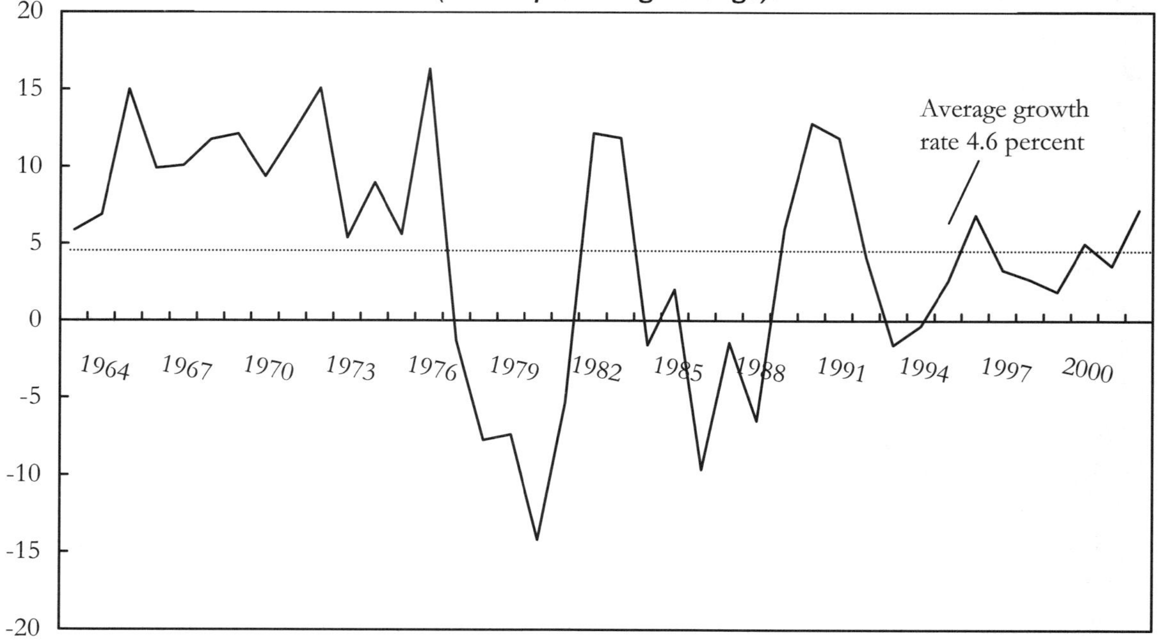

Sources: Iranian authorities; IMF staff estimates.

- During 1960–76, Iran enjoyed one of the fastest growth rates in the world: the economy grew at an average rate of 9.8 percent in real terms, and real per capita income grew by 7 percent on average. As a result, GDP at constant prices was almost five times higher in 1976 than in 1960. This stellar performance took place in an environment of relative political stability, low inflation (Figure 2), and improved terms of trade, as evidenced by the rising oil price relative to import prices (Figure 3). Both oil output and oil prices increased significantly during the period: oil production grew at an annual average rate of 10 percent, while oil prices relative to import prices increased by 214 percent during this subperiod.

- The growth trend was reversed during 1977–88, reflecting the turmoil in the aftermath of the 1979 revolution, the eight-year war with Iraq, the international isolation of Iran, the increased state dominance of the economy, and the plummeting of oil output and revenue. In 1988, oil production was only 36 percent of its 1976 level, and oil prices were 40 percent lower in real terms. This resulted in real GDP growth of −2.4 per year on average. Excluding oil output, non-oil GDP also declined, albeit at a more moderate pace (0.5 percent per year).

Figure 2. Inflation, 1960–2002

(Annual percentage change; end-period)

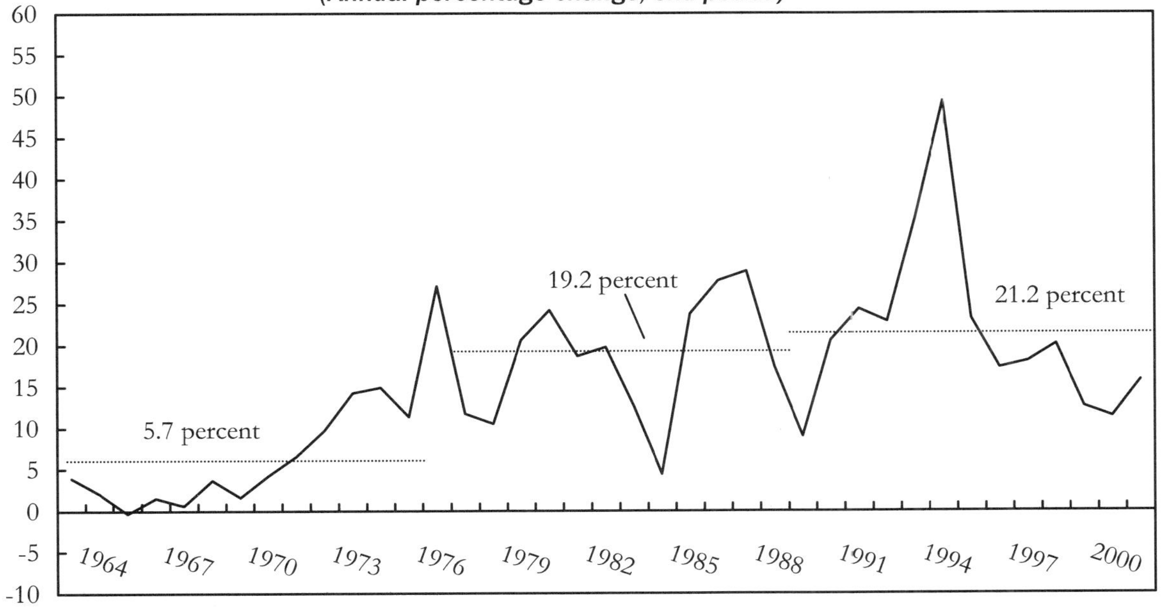

Sources: Iranian authorities; IMF staff estimates.

Figure 3. Oil GDP and Oil and Import Prices Ratio, 1960–2002

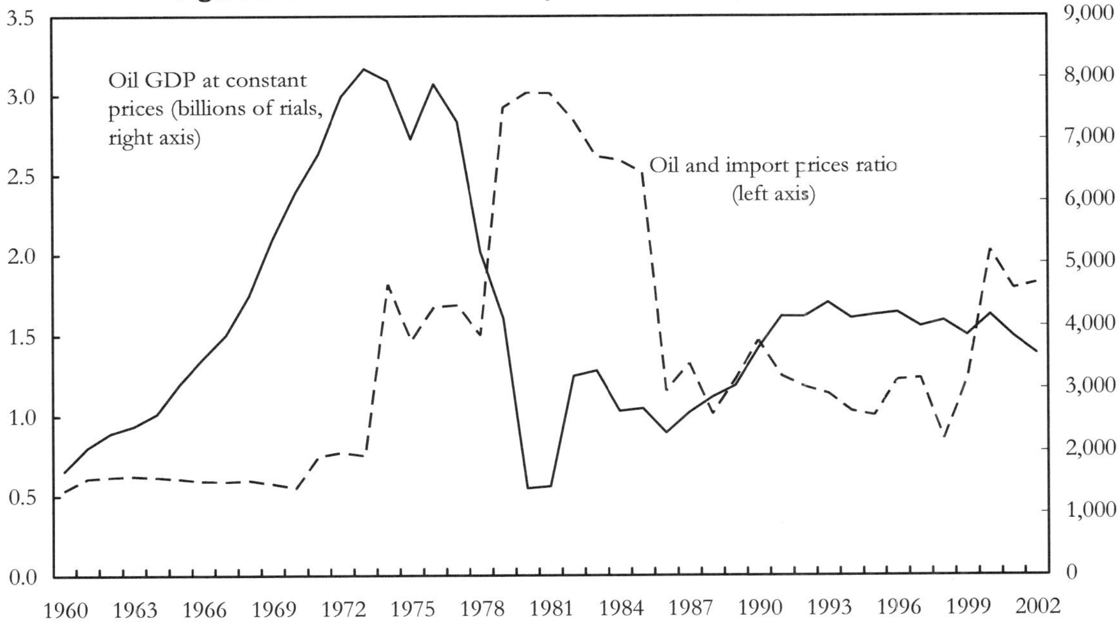

Sources: Iranian authorities; IMF staff estimates.

Table 1. MENA Region: Economic Growth, 1960–2002
(In percent, average for the period)

	1960–76	1977–88	1989–2002	1960–2002
Iran	9.8	–2.4	4.7	4.6
Oil-producing MENA countries[1]				
Algeria	5.9	2.3	2.0	3.5
Bahrain	4.5	4.3	4.7	4.5
Kuwait	4.1	–2.1	1.6	1.4
Libya	14.0	–0.2	–0.2	4.5
Oman	9.8	6.0	4.8	6.9
Qatar	7.0	–0.3	4.4	3.8
Saudi Arabia	7.7	2.0	2.5	4.2
United Arab Emirates	12.5	–0.5	6.2	5.1
Average, oil MENA countries	8.2	1.5	3.3	4.2
Non-oil MENA countries				
Egypt	4.6	5.7	3.3	4.5
Jordan	4.8	5.4	3.5	4.5
Lebanon	–3.9	5.2	0.3	0.3
Morocco	4.3	4.1	2.7	3.7
Pakistan	3.9	6.1	3.8	4.5
Sudan	2.9	2.4	4.3	3.2
Syrian Arab Republic	7.9	3.8	4.3	5.5
Tunisia	7.1	4.2	4.4	5.3
Yemen, Republic of	9.5	5.7	5.0	6.6
Average, non-oil MENA countries	4.6	4.7	3.5	4.2
MENA average (excluding Iran)	6.6	2.8	3.4	4.2

Source: IMF, International Financial Statistics (IFS).
[1]Excluding Iran.

- With the reconstruction effort and a partial recovery in oil output, real economic growth recovered during 1989–2002 to an average of 4.7 percent per year. This period, however, was marked by sharp fluctuations in the growth pattern, as the postwar economic boom (1989–93) was followed by the stagnation of 1993–94, when the economy was hit by lower oil prices, lack of external financing, and economic sanctions. The ensuing severe debt crisis, together with inappropriate macroeconomic policies, had an adverse effect on growth, which hovered around 3.6 percent from 1995 to 2000.

> During 2000–03, real GDP growth picked up to 6 percent as a result of significant progress in economic reforms—such as trade liberalization, exchange rate unification, an opening up to FDI, and financial sector liberalization—but also because of high oil prices and expansionary fiscal and monetary policies.

The growth performance of Iran compares favorably with that of the rest of the countries in the Middle East and North Africa (MENA) region, which averaged 4.2 percent per year during 1960–2002 (Table 1). Among the 17 countries in the region, only four—Oman, the Syrian Arab Republic, the United Arab Emirates, and the Republic of Yemen—grew faster than Iran. However, historical growth in Iran also exhibits higher variability than in the rest of the region.

Determinants of Economic Growth

The empirical studies on the determinants of growth can be broadly divided into two main categories. The first includes growth accounting exercises, which consist of estimating the contributions to growth of basic factor inputs—labor, physical capital, and human capital—and a residual that captures the efficiency at which physical and human capital resources are used, or TFP. The second comprises several empirical studies analyzing cross-country growth regressions to find the relationship between different explanatory variables and growth.

Growth Accounting

A standard growth accounting framework is used to discuss the historical sources of growth in Iran. We use the following Cobb-Douglas production function:

$$Y_t = A_t K_t^{\alpha} H_t^{1-\alpha}, \qquad (1)$$

where Y, K, and H represent output, physical capital, and human capital, respectively; α represents the contribution of physical capital to output; and t is an index for time. The term A represents TFP, or the efficiency at which the economy operates, which depends on factors such as the domestic political and international environment, the legal and regulatory framework, the creation and diffusion of more efficient technologies through international trade or FDI, and the effect of structural reforms, such as financial sector or labor market liberalization. Physical capital is considered as a homogeneous capital good, with no distinction made between equipment and nonequipment capital goods or between private and public capital goods (implicitly assuming that the productivity of the two types of capital is the same).[3]

To account for the effect of education on economic growth, a human capital index is constructed as a function of both the labor force and average years of schooling. However, in Iran it is difficult to measure the contribution of schooling to human capital because of the lack of an education quality index that would account for the changes in the productivity of education during 1960–2002.[4] Therefore, the paper considers two different specifications of human capital, which result in two different growth accounting exercises.

A basic specification is to make *human capital equal to raw labor*, that is, $H_t = L_t$. Under this specification, an increase in the average years of schooling of the labor force does not increase the productivity of labor. Given that the cross-country empirical evidence points to a positive effect of education on the productivity of labor, under this simple specification, the contribution of TFP to growth is overstated because it implicitly takes into account the effect of the quality of the labor input on output and growth.

A different assumption is to consider that schooling increases the productivity of the labor force along the following specification of human capital (Lucas, 1988):

$$H_t = L_t e_t , \qquad (2)$$

where L represents the labor force and e is the average years of schooling of the labor force.

The above-specified production function implies that *human capital accumulation exhibits increasing returns to scale*. This means that if we double both the number of workers and the average years of education for the labor force, human capital increases fourfold. Because anecdotal evidence—such as the increased proportion of college graduates with nonmarketable skills—points to a reduction in the quality of education in Iran over the period under study, the growth accounting exercise using this technology specification may result in an overstatement of the contribution to growth of human capital and an understatement of the contribution of TFP.

---

[3]This assumption is made because of the difficulty of measuring the productivity of public capital goods in Iran.

[4]Other proxies of the quality of human capital, such as an increase in productivity of workers (measured by their relative salaries, presumably reflecting relative educational attainment), are not available in Iran.

Table 2. Sources of Economic Growth (Raw Labor), 1960–2002
(In percentage points)

	Average Growth Rate	Contribution of Capital	Contribution of Raw Labor	Contribution of TFP
1960–76	9.8	3.9	1.2	4.7
1977–88	–2.4	1.7	1.4	–5.5
1989–2002	4.7	2.3	1.5	1.0
1960–2002	4.6	2.1	1.4	1.1

Sources: Iranian authorities; IMF staff estimates.

Taking natural logarithms and differentiating with respect to time, the following decomposition of growth is obtained:

$$g_Y = g_A + \alpha g_K + (1-\alpha) g_H , \qquad (3)$$

where g denotes the growth rate of the variable in the subscript. If factor markets are competitive, the first-order profit-maximizing conditions for the firm imply that α corresponds to the share of rental payments to capital in total income, that is, $\alpha = (r - d)*K/Y$, where r is the net rate of return on capital and d is the depreciation rate of capital. The derivation of the time series of the capital stock and the data sources are presented in Appendix 1.

The results of the growth accounting exercises under the two alternative specifications for human capital ($H_t = L_t$ and $H_t = L_t e_t$) are shown in Tables 2 and 3, respectively.

The noninclusion of the effect of increased schooling on the productivity of the labor force and on growth results in a positive contribution of TFP to growth during 1960–2002, because changes in the quality of the labor force are implicitly included in TFP. Under the alternative specification, which considers that human capital increases linearly with average years of schooling, the growth accounting exercise suggests that the contribution of

Table 3. Sources of Economic Growth (Education), 1960–2002
(In percentage points)

	Average Growth Rate	Contribution of Capital	Contribution of Human Capital	Contribution of TFP
1960–76	9.8	3.9	2.7	3.2
1977–88	–2.4	1.7	5.5	–9.6
1989–2002	4.7	2.3	4.3	–1.8
1960–2002	4.6	2.1	3.7	–1.2

Sources: Iranian authorities; IMF staff estimates.

TFP to growth becomes negative (–1.2 percent on average during 1960–2002) (Table 3).

Under both accounting exercises, the contribution of TFP to growth is positive during the high-growth subperiod of 1960–76 and becomes negative during the political turmoil and war period of 1977–88. This result points to the critical importance of political and external developments for Iran's economic growth.

The results of the two exercises differ for 1989–2002. Under the first specification, in which human capital equals raw labor ($H_t = L_t$), the contribution of TFP to growth is positive; however, if we consider a linear effect of education on human capital ($H_t = L_t e_t$), the contribution of TFP to growth becomes negative. A more realistic TFP estimate may lie between these two extreme cases. In particular, it is likely that there was a very small (or even a negative) contribution of TFP to growth during the 1989–2002 subperiod,[5] owing to slow progress in structural reforms and an increase in both inflation and its volatility.

[5]The TFP growth estimates are subject to measurement errors of physical and human capital. However, using the same growth accounting methodology, the TFP growth estimates for Iran are systematically lower than those in most other developing countries.

Empirical Analysis

Cross-Country Empirical Evidence

Some stylized facts from cross-country growth regressions are relevant for Iran:

- There was a positive relationship between the level of education of the labor force and economic growth. Barro (1991 and 1997) and Benhabib and Spiegel (1994) show that the initial level of education is an important factor in explaining subsequent growth. However, Bils and Klenow (2000) find that the causality goes from growth to increases in school enrollment rates.

- Macroeconomic stability—usually defined as a combination of a low inflation rate, low budget deficits, and undistorted foreign exchange markets—improves the business environment and reduces the uncertainty on the return of investment projects and, therefore, has a positive relationship with economic growth. Fischer (1993) and Bleaney (1996) find that macroeconomic instability (measured by a combination of high inflation, fiscal imbalances, and high volatility of the real exchange rate) has a significant negative effect on economic growth and possibly also a negative effect on investment. More specifically, the literature finds a negative relationship between inflation and growth. Khan and Senhadji (2001) find that an inflation rate above 11–12 percent is associated with a significant reduction in growth in developing countries. Sarel (1996) finds that high inflation—above 8 percent per year—has a negative and statistically significant effect on growth, and that doubling the inflation rate would reduce the average growth rate by 1.7 percentage points. Barro (1997) finds a smaller effect: an increase in the inflation rate of 10 percentage points would reduce the growth rate by 0.2–0.3 percentage points.

- Financial development reduces the cost of capital and generally has a positive correlation with growth, but the direction of the causality is difficult to establish. Demetriades and Hussein (1996) studied the experience of 16 countries: 4 displayed causality from financial depth to growth, 4 displayed causality from growth to financial depth, and 7 displayed a feedback relationship between finance and growth. Regarding financial repression and growth, both Roubini and Sala-i-Martin (1995) and Arestis and Demetriades (1997) find a negative relationship between financial repression indicators and growth. The only exception was Korea, in which financial repression favored the growth-leading export sector.

- Trade openness generates technology spillovers and provides the economy with access to specialized inputs from abroad. The literature finds a significant effect of trade openness on growth. Greenaway, Morgan, and Wright (1998) cover 73 countries and use a dynamic regression framework to investigate potential lagged effects of openness on growth. They find that the positive effect of trade openness on growth becomes more significant over the long term, whereas in the short term this effect is much less important. Improvements in the terms of trade are generally associated with faster growth (see Barro, 1996 and 1997; Easterly and others, 1993; and Fischer, 1993).

- Finally, other factors, such as political variables and income inequality, may also play an important role in economic growth. Alesina and others (1996) find a significant negative relationship between political instability and growth. This result is particularly strong when there are significant changes in the ideological position of the executive branch. In another empirical study, Mauro (1995) finds a negative correlation between political risk and economic growth. Other empirical studies show mixed results on the relationship between income inequality and economic growth.

Empirical Analysis of Factors Affecting Non-Oil GDP Growth

This section focuses on the empirical relationship between non-oil GDP growth and some factors commonly referred to in the literature as having a significant effect on growth performance—namely, trade openness, macroeconomic stability, terms of trade changes, and financial development. Given the importance of changes in oil production and political developments in Iran, the empirical analysis also includes these two variables.

To examine the link between trade openness and economic growth, we adopt the ratio of imports to non-oil GDP as a proxy for trade openness because of the lack of data on average tariffs and nontariff barriers to trade for the entire period 1960–2002. Macroeconomic stability is proxied by the inflation rate, owing to the lack of data on government deficits and exchange rate distortions during 1960–2002. Terms of trade are proxied by the change in the ratio of oil prices to import prices of industrial goods. Because oil represents about 80 percent of Iran's exports, and 95 percent of imports are industrial goods, the ratio of the prices of these two types of commodities is a good proxy of the terms of trade. Financial development is proxied by the change in the ratio of broad money (M2) to non-oil GDP. Finally, we include a dummy variable for the subperiod 1977–88 to take into account the effect of political instability and war on growth.

Table 4. Non-Oil Annual GDP Growth, 1961–2002

Explanatory Variables	Coefficient	*T-Value*
Constant	4.28002	2.3848
Change in oil production (–1)	0.05611	2.6605
Imports/GDP ratio	0.32252	6.3954
Consumer price index	–0.31006	4.7403
Oil imports/price ratio	0.04794	2.9851
M2/GDP ratio	–0.27408	3.4136
Dummy 1977/88	–5.98764	5.3164
R^2	0.82758	
Tests	**Value**	***P-Value***
Chow (1982:1)	0.4842	0.935
Chow (1998:1)	0.5893	0.673
AR 1–4 test	1.1225	0.364
ARCH 1–4 test	0.1132	0.977
Hetero test	4.6767	0.946

Source: IMF staff estimates.

Using the econometric package PcGets, which allows for an automatic reduction of a general model to a parsimonious one, we establish an empirical relationship between non-oil GDP growth and its explanatory variables, shown in Table 4.

The *t*-statistics of the regression show that all variables are significant at the 95 percent confidence level and explain 82.8 percent of variance of growth; there are no structural changes during the period (Chow tests); no autocorrelation (AR test); and no heteroscedasticity (ARCH and hetero tests) of the residuals of the regression.

The above results are largely consistent with the cross-country evidence on economic growth. Increased trade openness and macroeconomic stability (measured by inflation rates) are positively correlated with growth. Also, improvements in the terms of trade are positively correlated with growth. Higher oil production stimulates non-oil GDP growth through higher demand of inputs from the non-oil sector and because higher oil revenues stimulate public expenditures, particularly public capital expenditures

Table 5. Average Sectoral Growth, 1960–2002
(In percent, in real terms)

	1960–76	1977–88	1989–2002	1960–2002
Agriculture	4.6	3.9	4.1	4.2
Oil and gas	10.0	–8.6	2.5	2.4
Industries and mines	14.0	–1.3	7.3	7.6
Services	11.1	–1.9	4.8	5.4
Non-oil GDP	10.1	–0.5	5.0	5.5
GDP	9.8	–2.4	4.7	4.6

Source: Iranian authorities.

(see Chapter 5 on oil revenue management). Of all the variables studied here, political instability and war have the strongest (negative) effect on growth, reducing growth by about 6 percentage points per year during the 1977–88 subperiod.

The main difference from the cross-country evidence in other studies concerns the negative effect of financial deepening (measured by the change in the ratio of M2 to non-oil GDP) on growth in Iran.[6] The lack of positive correlation between financial deepening and growth in Iran, which is in contrast with other countries, could be attributed to an inefficient financial system that channels resources to investments with very low productivity or to the sectors with lower growth potential. The low or even negative TFP contribution to growth in Iran is likely to have captured these financial system deficiencies as well.

Sectoral Growth

During 1960–2002, the industrial sector exhibited the strongest performance. Industrial output (mostly manufacturing and construction) grew at 7.6 percent per year on average and was 23 times higher in 2002 than in 1960. As a consequence, the share of industrial output to GDP increased

[6]With alternative specifications of the model, such as using changes in real money, the relationship between financial deepening and growth becomes statistically insignificant.

during this period from 7 percent to 25 percent. In contrast, the oil sector grew by just 2.5 times, and its relative weight decreased from one-third of GDP at constant prices in 1960 to less than 13 percent in 2002. The output in the agricultural sector grew at 4.2 percent on average, a slightly slower pace than GDP, but well above the average population growth of 2.6 percent, whereas the services sector grew at a faster pace than GDP (5.4 percent per year on average during the period). Table 5 shows the average growth rates for each sector.

Despite the rapid growth of the industrial sector, the low or even negative growth of TFP during the period under study—together with high physical capital investment—suggests a low productivity of investment in the industrial sector, possibly reflecting trade restrictions[7] and inefficient public sector investment in the industrial sector.

Summary of Empirical Findings

The analysis covers a period of 42 years (1960–2002) that witnessed significant political and social changes as well as periods of instability (for example, the 1979 revolution, the war with Iraq, economic sanctions). The analysis in the previous sections shows the following:

- Changes in the political environment have had a major impact on economic growth in Iran. The periods of relative political stability and absence of major external conflicts (1960–76 and 1989–2002) are clearly associated with high GDP growth, whereas the political turmoil and war period of 1977–88 was associated with negative growth. The paper shows that the average annual growth rate during the 1977–88 period was reduced by 6 percentage points as a result of these factors.

- The fivefold improvement in the average level of education of the labor force since 1960 may explain up to half of total economic growth in the past 42 years.[8] However, it is difficult to determine precisely the magnitude of the contribution of investment in education to growth owing to a lack of the data needed to measure the effect of educational attainment on productivity.

[7]Despite recent reductions in import taxes and nontax barriers, Iran's trade regime is very restrictive: the average (unweighted) tariff rate in 2002 was 30 percent, the eleventh highest tariff rate out of 193 surveyed countries, according to the Trade Restrictiveness Index Ratings in the Trade Policy Information Database (TPID) maintained by the Policy Development and Review Department, IMF.

[8]The quantification of the contribution of education to growth depends on the specification of the production function.

- Trade openness is significantly associated with faster GDP growth. An increase in the imports-to-GDP ratio of 1 percentage point is associated with faster growth of 0.3 percentage points.

- Regarding macroeconomic stability and growth, there is a positive link between growth and lower inflation in Iran, and this relationship is statistically significant. A reduction in the inflation rate of 1 percentage point would increase potential growth by 0.3 percentage points.

- Given the inefficiencies in the financial sector, the link between financial deepening and growth in Iran is not clear. In fact, when financial development is proxied by changes in the ratio of M2 to non-oil GDP, the relationship between these two variables becomes negative. Given the cross-country empirical evidence of a generally positive relationship between financial development and growth, it is likely that changes in the financial system that would increase its efficiency would yield potentially large gains in terms of long-term growth, reversing the observed negative relationship between financial depth and growth in Iran.

Policy Lessons

Three main policy lessons can be derived from the Iranian growth experience:

- Structural reforms in a stable political environment are key to improving growth performance over the medium and long term. To increase the long-term growth rate of the economy above its historical mean of 4.6 percent per year, policies should be directed at increasing productivity (measured by TFP). Moreover, the cross-country empirical evidence and the empirical findings for Iran show that growth is directly associated with factors such as trade openness, macroeconomic stability, and political stability. These findings call for stepped-up implementation of structural reforms—trade and FDI liberalization, privatization and deregulation to increase the size and role of the private sector, and further financial sector reform to eliminate financial repression that harms long-term growth. Other reforms, such as the elimination of subsidies and the implementation of fiscal, monetary, and exchange rate policies aimed at increasing macroeconomic stability, would also play a critical role in enhancing growth performance.

- Increases in the efficiency of human capital resources through investment in education appear to be an important reason for Iran's

Table 6. Comparison of Investment and Growth Performance of Iran with Six High-Growth Asian Economies, 1962–2002
(In percent)

	Average GDP Growth	Average Investment/GDP	Investment/GDP Growth
Growth			
Hong Kong SAR	7.0	26.1	3.7
Indonesia	5.4	21.7	4.0
Korea	7.8	27.9	3.6
Malaysia	6.7	28.6	4.3
Singapore	7.9	36.0	4.5
Thailand	6.7	26.5	4.0
Average	**6.9**	**27.8**	**4.0**
Iran	4.6	30.5	6.6

Sources: IMF, IFS; IMF staff estimates.

growth. In this respect, achievements in Iran since the 1979 revolution have been very important: the average level of schooling for the working population has more than tripled (from 1.5 years of schooling in 1979 to about 5 years in 2002). Education policies aimed at allocating increased resources to primary and secondary education and promoting on-the-job training programs would, in all likelihood, further enhance growth prospects. The need for further efforts in education is evident when we consider that Iran has an illiteracy rate of about 20 percent, despite the substantial progress since 1979.

- Finally, with respect to the contribution of physical capital to economic growth, Iran's investment rate (which averaged more than 30 percent during 1960–2002) is already high by international standards, even when compared with the high-growth countries of East Asia (see Table 6). Its payoff, however, as measured by average incremental capital-output ratios, does not suggest that it should be increased further, but that the efficiency of investment projects needs to be improved. The low efficiency of many investment projects undertaken in Iran, especially in agriculture, industry and mining, and housing, could be explained in part by subsidized energy and inputs and negative real interest rates on bank financing. Nonetheless, despite the high rates of investment over the past years, physical infrastructure is in need of upgrading and modernization.

CHAPTER

2 The Financial Sector: Issues and Reforms

The Iranian financial system has evolved through a number of stages since the 1979 revolution. After widespread nationalization in the 1980s, reform of the financial system in the 1990s and 2000s focused on improving the regulatory environment and streamlining controls to enhance efficiency, while more limited steps were taken to open the system to private sector participation and foster competition. These efforts, however, have not significantly altered the structure of the financial system, which remains underdeveloped and exhibits several weaknesses that are typical of countries in transition to a market economy. These weaknesses include ownership of and dominance over financial institutions by the public sector (even though a number of private banks have started operations in recent years), widespread use of administrative controls on credit allocation and rates of return (equivalent to interest rates), a lack of competition among banks, relatively weak bank supervision, shallow and weakly regulated capital markets, and an underdeveloped insurance industry (Table 7). Financial system reform must continue in line with the gradual opening up of the economy to foreign trade and capital inflows, the increasing role of the private sector in economic activity, and the need to enhance bank supervision and improve monetary policymaking.

State-owned banks, including the specialized banks, continue to be the major providers of financing to the corporate sector, even though their lending is rationed and highly concentrated on a small number of large companies or priority sectors (notably agriculture) to which they lend at subsidized rates. Despite its increasing capitalization (24 percent of GDP), the stock market has a relatively low level of liquidity and provides little fresh financing to the corporate sector.

Note: This chapter draws on the IMF's Financial Sector Assessment Program (FSAP) undertaken in 2000 and provides updated information and analyses on several aspects of the Iranian financial system. It was prepared by Vitali Kramarenko.

Table 7. Comparative Financial Development Indicators, 2000–01
(Comprehensive index, scale 0 –10)

	Comprehensive Index	Banking Sector	Nonbank Financial Sector	Regulation and Supervision	Monetary Sector and Policy	Financial Openness	Institutional Environment
Iran	2.2	2.5	3.0	3.7	0.6	0.0	3.7
MENA average	5.4	5.3	4.8	6.5	5.4	6.1	4.7
Finacial development levels (average scores)[1]							
High	7.5	7.3	6.7	8.9	7.3	8.9	5.9
Medium	5.3	5.0	4.1	6.5	5.6	6.1	4.8
Low	3.3	3.1	2.7	3.5	3.1	3.9	3.8

Source: Creane and others (2003).
[1]Within overall scale of 0–10, intermediate scales are as follows: high, above 6; medium, 4–6; and low, below 4.

The financial system has operated in a volatile macroeconomic environment. Between 1989/90 and 2003/04, real GDP growth varied considerably around an average of about 5 percent, with standard deviation amounting to 3¾ percentage points. Also, average annual CPI inflation fluctuated in a range of 11–49 percent, and the official exchange rate vis-à-vis the dollar depreciated from Rls 80 to Rls 8,400. More recently, large unsterilized purchases of foreign exchange and the relaxation of credit policy led to high rates of growth of credit and money (with annual average growth rates of 36 and 29 percent, respectively) in the three years leading into 2003/04. This led to the persistence of inflation at 15½ percent for the last two of those years.

The Iranian financial system lags behind those of other MENA countries in many respects. According to an overall index of financial development prepared by the IMF staff (see Creane and others, 2003), Iran ranks low among MENA countries, with particularly low scores for financial openness, monetary policymaking, and development of the banking and nonbanking financial sectors (see Table 7).[1]

This chapter reviews key issues and reform challenges in the financial sector in Iran. The following section provides information on the evolution of the size and structure of the banking system, capital markets, insurance sector,

[1]Monetary policy issues are discussed in Chapter 3.

and foreign exchange market. The section on governance and regulatory oversight discusses issues of regulation and governance of Iran's financial system and highlights the reasons behind the relatively low level of financial intermediation. It also provides an update on the implementation of FSAP recommendations. The section on the reform agenda presents links between financial sector reforms and the performance of the real economy, drawing on cross-country experience, and outlines the reform agenda in selected areas based on earlier FSAP recommendations. The last section in this chapter summarizes and offers some recommendations.

Overview of the Financial System

Banks and Nonbank Credit Institutions

Following the 1979 revolution, all commercial banks were nationalized, and foreign participation in banking was banned. At the time of nationalization, the banking network included 36 banks with various degrees of government and foreign ownership, comprising 7 specialized banks, 26 commercial banks, and 3 regional financial institutions. Only since 2001 have private banks been reauthorized to operate in Iran.

The structure of the banking system has not changed substantially since the nationalization, which reflects the continued dominance of the public sector in the economy in general. Currently, the banking system consists of six state-owned commercial banks, four state-owned specialized banks (see Box 1), a state-owned Postal Bank (licensed in 2004), six recently established small private banks, and one nonbank private credit institution. State-owned commercial and specialized banks dominate the banking system, holding about 98 percent of deposits. At end-March 2004, the consolidated assets of banks amounted to 49 percent of GDP, and broad money, excluding foreign currency deposits, represented 45 percent of GDP.[2] These ratios are below those recorded in the 1990s and those of most comparable MENA countries (Figure 4). Lack of progress in deepening financial intermediation is largely attributable to high inflation (about 20 percent on average over the 10 years ending in 2004) and various administrative controls on banking operations (see the following section).

The financial position of the banking system is relatively weak. Despite a recapitalization of state-owned banks in 2002 amounting to Rls 5,000 billion (0.7 percent of GDP) and the reinvestment of postunification foreign

---

2These ratios do not include the Postal Bank and some other banking institutions, which are not covered by the monetary survey.

Box 1. Specialized Banks

There are four specialized banks: housing (Bank Maskan), agriculture (Bank Keshavarzi), export development (Export Development Bank), and industry and mining (Bank Sanat-O-Madan). These banks also take deposits, but the great majority of their loanable funds comes from commercial banks, the central bank, and other public sources, including the central government.

Bank Keshavarzi (Agricultural Bank) is by far the largest of the specialized banks; it is comparable in size to other state-owned banks, except Bank Melli. Its loan portfolio amounted to Rls 29 trillion (2.5 percent of GDP) as of end-March 2004, accounting for about 60 percent of banking system loans to agriculture. The resource base is composed of government contributions, central bank loans, deposits of other banks, and a growing number of deposits by nonbanks. Bank Keshavarzi offers a wide range of Islamic finance instruments for agricultural financing and has set up an Agricultural Insurance Fund covering 63 commodities. Such an insurance system is more common in developed countries and rare in developing countries because of the high risks associated with agricultural activity. However, the government has provided large financial support to the bank to compensate for drought-related losses.

exchange gains in equity capital of state-owned banks in 2004, the risk-weighted capital adequacy ratio, at 7.2 percent, is below the 8 percent recommended by the Capital Adequacy Framework in the Basel Capital Accord. In 2003/04, the return on assets was estimated at 1 percent, and the ratio of nonperforming loans was reportedly 5.2 percent (Table 8).[3] Private banks, however, have a much stronger financial position than implied by the banking system's average ratios.

The degree of concentration in the banking system remains high, with the state-owned Bank Melli controlling about one-third of assets. There is also substantial loan concentration in individual state-owned banks. On a weighted average basis, the 20 largest exposures of each of the state-owned banks accounted for 24.3 percent of their total committed financing facilities in 2000. This suggests a lack of diversification in the asset portfolios of banks. Furthermore, following the exchange rate unification in March 2002, the extent of dollarization of the banking system has increased, albeit from a low base.

[3]These ratios are not fully comparable to those in other countries owing to differences between Iran's accounting standards and the International Accounting Standards, as well as a lack of proper regulations on loan classification and provisioning.

Figure 4. Selected MENA Countries: Broad Money/GDP Ratio, 1990–2002
(In percent)

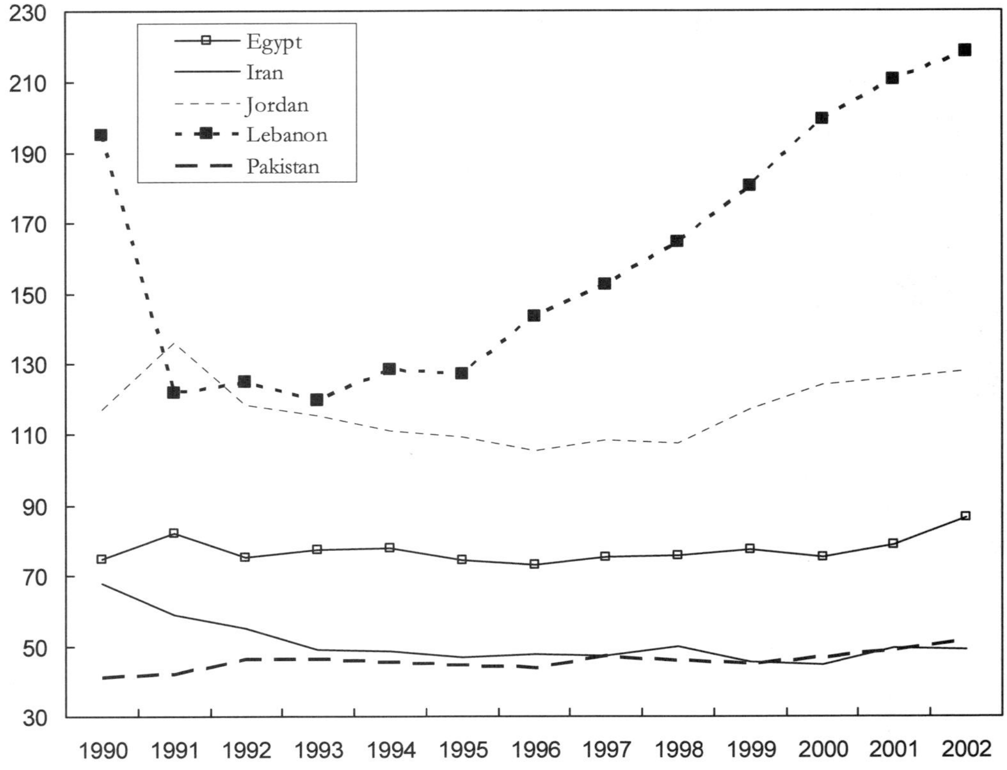

Source: IMF, World Economic Outlook (WEO) database.

The recently licensed private banks, which emerged from private nonbank credit institutions that were authorized in the mid-1990s, are mainly active in some market niches, such as short-term bridge financing of medium-size enterprises,[4] mortgage lending, and retail consumer lending. Private banks have started to exert competitive pressure on state-owned banks.

Iran's financial system also includes a number of small banking institutions and informal financial intermediaries. About 6,000 *Qarz-ul-Hasanah* funds, which raise zero-interest funds and provide interest-free loans, operate on a small scale. Almost 1,000 registered credit cooperatives are also in operation, but these are very small, with total assets well below Rls 1,000 billion

[4]Because there are long waiting lists for loan applications in state-owned banks, many companies in urgent need of liquidity apply for bridge loans from private banks; these loans are subsequently refinanced by state-owned banks.

Table 8. Financial Soundness Indicators, 2003/04
(In percent)

	Banking System	State-Owned Commercial Banks	State-Owned Specialized Banks	Private Banks
Risk-weighted capital adequacy of banks	7.2	5.5	15.4	19.5
Ratio of nonperforming loans of banks	5.2	4.7	8.5	2.1
Return on average assets of banks	1.0	0.6	3.0	3.7
Return on equity	20.4	16.7	23.9	35.0
Net open position in foreign exchange to capital	30.0	40.0	14.0	9.0

Source: Iranian authorities.

(0.1 percent of GDP). Moreover, some *bonyads* (charitable foundations) run quasi-banking institutions, one of which has recently received a credit institution license. Finally, informal finance is common with high rates of return, reflecting lack of access to bank financing by small and medium-size enterprises.

Capital Markets and Insurance

Following the 1979 revolution, the activities of the Tehran Stock Exchange (TSE) came to a standstill. The TSE was reopened in 1989 when the government listed many state-owned companies in an effort to start a divestiture program and develop the private sector in the context of the First Five-Year Development Plan. Since then, TSE activity has continued to grow, despite some volatility that reflected uneven progress in macroeconomic stabilization and structural reforms, as well as the variations in the price of oil.

Since 1998, the size of the capital market has been increasing rapidly. By end-April 2004, the capitalization of the TSE reached $37.5 billion (24 percent of GDP), mainly owing to a more than 600 percent increase in the TSE share price index since the last downturn of late 1998, but also owing to the increased number of listed securities (Figure 5). Capitalization of the TSE relative to GDP, however, remains below that of comparable exchanges in Egypt and Jordan. Despite the increase in capitalization, little fresh financing has been provided to the corporate sector.

Figure 5. Selected MENA Countries: Share Price Indices, 1998–2003
(1998 = 100)

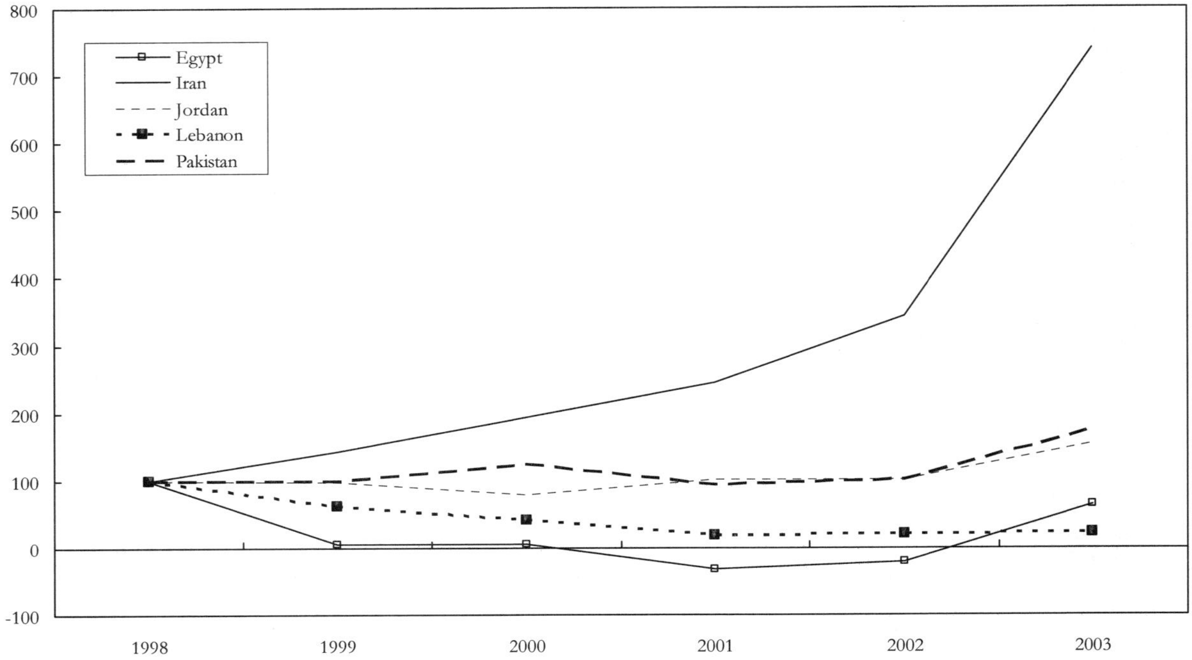

Sources: Iranian authorities; IMF staff estimates.

The recent increase in the stock market's capitalization has been driven by a number of factors. First, the market is recovering from the low level of 1998, when the average price-earnings (P/E) ratio was about 3. Second, business confidence has strengthened as a result of recent reforms and liberalization measures, rapid economic growth, and high oil prices, thereby significantly improving investors' expectations of future profit growth. Third, a relaxation in the monetary policy stance during 2002–04 may have contributed to portfolio adjustment away from bank deposits, which carry negative real rates of return, in favor of investment in the stock market. Fourth, anecdotal evidence suggests that unrecorded portfolio investments from Iranians living abroad also may have contributed to the stock market rally. Finally, the opening of new regional branches of the TSE has attracted a growing number of investors from the provinces, increasing demand for equity investment. Although the average P/E ratio is still moderate compared with those in other equity markets (about 9), its level might be understated given possible weaknesses in accounting and reporting rules (Table 9).

Table 9. Selected Emerging Markets: Price-Earnings and Turnover Ratios, April 2004

Market	Price-Earnings Ratio	Turnover Ratio
Egypt	13.3	1.5
Indonesia	9.6	4.2
Iran	9.0	0.2
Jordan	19.8	1.3
Malaysia	21.7	2.8
Morocco	16.7	1.1
Russia	17.0	4.9
Turkey	10.0	17.5

Sources: International Finance Corporation; Tehran Stock Exchange.

Most listed companies are parastatal enterprises with varying degrees of direct and indirect government ownership. The ongoing divestment program has been modest so far (1 percent of GDP per year).[5] The largest sectors in TSE capitalization are automotives, mining, petrochemicals, and financial services (primarily intermediaries).

Despite the stock market rally in recent years, market turnover remains low. At about 0.2 for 2003/04, it is substantially lower than the turnover ratios for other emerging markets (Table 9). This reflects the relatively large presence of a few institutional investors that do not engage in active trading. A large share of capitalization is reportedly accounted for by cross-shareholdings of some companies and insufficiently regulated investment companies.

The insurance sector is still very small. The Central Insurance Authority (Bimeh Markazi) is regulator and supervisor, as well as a market participant in its own right through reinsurance. Five companies owned directly or indirectly by the government collected about 1 percent of GDP in premiums in 2002/03. The recent authorization and licensing of private insurance companies is expected to enhance the development of this sector.

[5]The divestment program mainly offers large blocks of shares to strategic institutional investors.

Foreign Exchange Market

After years of tight restrictions, Iran has largely liberalized current account transactions and made progress in trade liberalization. On March 21, 2002, the exchange rate was unified, most exchange restrictions on current account transactions were eliminated, and import-related transactions in the financial system were liberalized.[6] At present, there are no derivative instruments to hedge against exchange rate risk, and all transactions are carried out in the spot market. The central bank remains the main seller in the domestic and offshore foreign exchange markets (mainly the Kish Island market).

Some capital account liberalization measures have recently been taken. A new law approved in 2002 established a clear legal framework for FDI in Iran and contributed to the increase in FDI commitments, excluding oil and gas, to $1.8 billion in 2003/04 from about $70 million in 2001/02. Other forms of capital inflows are subject to restrictions—mainly through limitations on non-Iranian nonresidents' investment in the stock market and real estate. It appears that nonresident Iranian nationals have recently increased substantially their portfolio investment in Iran; this is legally authorized but not statistically recorded.

Regarding outflows, following the exchange rate unification, two important avenues have been opened for legal (but largely unregulated) access to foreign exchange for current and capital account transactions. First, unregulated transfers of rials can be made to offshore zones (subject to compliance with anti–money laundering (AML) regulations), where they can be converted into foreign exchange without restrictions. Second, the use of foreign exchange initially originating from export proceeds, short-term capital inflows in the form of banking deposits, and remittances of Iranians abroad is not subject to restrictions in the offshore or onshore markets.

Governance and Regulatory Oversight

The Iranian financial system operates under Islamic finance principles (i.e., consistent with Sharia) based on the 1983 Law on Usury-Free Banking. Under these principles, ex ante preset interest rates are prohibited and the return on financial instruments must be linked to purchase and resale of goods (and services) or to profit and loss sharing on projects. However, in practice, little uncertainty exists on future rates of return in Iran. Commercial banks achieve this by smoothing the returns (that is, implicitly building up and drawing down reserves to equalize returns over time) and through

[6] A more detailed historical account of the evolution of the exchange rate regime is provided in Chapter 4.

Box 2. Financial Repression: Definitions and Cross-Country Evidence

Financial repression is defined as the set of policies, laws, regulations, taxes, qualitative and quantitative restrictions, and controls imposed by the government that do not allow financial intermediaries to operate at their full potential (Roubini and Sala-i-Martin, 1995). These public policies artificially increase demand for base money of banks (for example, through high required reserves) and households (for example, via constraints on financial innovations and the imposition of ceilings on a range of return-earning instruments). Given this artificial increase in demand for base money, revenue from inflation tax is increased. This revenue is "spent" indirectly through subsidized directed credits and directly via financing of the government. Even though it is possible in the short run to generate a higher inflation tax under financial repression than under a liberal financial system, the many distortions and inefficiencies created by financial repression impede growth in the medium and long term. The empirical research on the relationship between financial repression and growth (Roubini and Sala-i-Martin, 1992; Levine, 1997; and Demetriades, Devereux, and Luintel, 1998) suggests that financial repression affects growth in three ways. First, the productivity of investment is reduced (low and declining TFP). Second, overall savings and investment are lower. Third, intermediation costs of allocating savings to investment are higher. The last two effects are associated with disintermediation and financial system weakening.

implicit or explicit government guarantees on returns and principal of financial instruments issued by state-owned banks.

The Iranian financial system is subject to "financial repression" (Box 2), through various forms of control, in particular on the banking system. These controls have contributed to low profitability and undercapitalization of state-owned banks. The banking system in its present form largely fulfills quasi-fiscal functions in the context of a public-sector-dominated economy. It channels financial resources to priority sectors as defined by the government rather than to the projects with the best risk-return opportunities. Moreover, full transparency in relations between state-owned banks on the one hand and public enterprises, charitable foundations (bonyads), and influential large private companies on the other appears to be lacking. Anecdotal evidence suggests that this lack of transparency manifests itself in long waiting lists and rationing in loan applications at negative real rates of return.

Low profitability and undercapitalization of state-owned banks are interconnected problems. The former mainly stems from administrative controls on rates of return on deposits and loans (Tables 10 and 11), sectoral credit allocation, directed credits, high reserve requirements, government interference in management, and high operating costs. Some forms of administrative controls have recently been eased, in particular those related to sectoral credit allocation, but other controls continue to hinder competition and profitability. In turn, low profitability contributes to a slow buildup of equity capital.

Table 10. Rates of Return on Deposits, 1999/2000–2003/04[1]
(In percent per year)

	1999/2000	2000/01	2001/02[2]	2002/03	2003/04
Short-term	8.0	8.0	7.0	7.0	7.0
Long-term					
One-year[3]	14.0	14.0	13.0	13.0	13.0
Two-year[3]	15.0	15.0	13–17	13–17	13–17
Three-year[3]	16.0	16.0	13–17	13–17	13–17
Four-year[3]	. . .	17.0	13–17	13–17	13–17
Five-year[2]	18.5	18.5	17.0	17.0	17.0

Source: Iranian authorities.
[1]Iranian fiscal year ends March 20.
[2]Rates effective from May 12, 2001.
[3]Long-term deposits over one year introduced in 1990/91 and 2000/01.

The four private commercial banks are not subject to controls on rates of return and do not benefit from implicit guarantees of deposits. As a result, their costs of funds, including deposit rates, are higher, which tends to be reflected in higher lending rates compared with those of state-owned banks. Despite this pricing disadvantage, private banks have been able to increase their market share, owing to their superior customer services, including faster processing of applications and more customer-tailored banking products, and to credit rationing by state-owned banks.

Banking supervision is undergoing major changes but is still focused on compliance with government directives rather than risk assessment. The central bank is in charge of supervising banks and large credit institutions. However, small nonbank credit institutions, including credit unions and Qarz-ul-Hasanah, are not subject to supervision by either the central bank or the ministry of finance and economy. These institutions have been authorized by the ministry of interior, which is also responsible for their oversight and that of other nonprofit organizations. Legislation bringing these institutions under central bank supervision is under consideration.

A comprehensive program to develop and implement a risk-based regulatory and supervisory framework for the banking sector is currently under way, in line with FSAP recommendations. Some regulatory reforms are in place, including licensing, net open positions in foreign exchange, the definition of statutory capital, capital adequacy, large exposures, connected lending, and

Table 11. Rates of Charges on Bank Loans, 1999/2000–2003/04[1,2]

(In percent per year)

	1999/2000	2000/01	2001/02[3]	2002/03	2003/04
Agriculture	13–16	13–16	14–15	13–14	13.5
Industry and mining	17–19	17–19	16–18	16	16.0
Housing	15–16	15–16	15–16[4]	14–15[4]	15.0[4]
	18–19	18–19	17–19	16–18	18.0
Trade and services	22–25	22–25	23[5]	22[5]	21.0[5]
Export	18	18	18	17	15.0

Source: Iranian authorities.

[1]Iranian fiscal year ends March 20.
[2]Announced rates representing minimum payable return. May be lower or higher than the actual ex post rates of return.
[3]Rates effective from May 12, 2001.
[4]Only for Bank Maskan (housing bank).
[5]Minimum rate.

AML regulations for banks. Supervisory functions have been unified under a single department at the central bank. Onsite and offsite inspections have begun, using risk-based criteria. Finally, reporting forms and supervision manuals are being developed. Despite this progress, a full-fledged, risk-based supervision framework has not been established yet, and the supervision of state-owned banks continues to rely on tight monitoring of credit allocation and compliance with administrative restrictions.

Regulatory oversight of publicly traded securities and stock exchange operations is relatively underdeveloped. The TSE operates according to the 1966 Stock Exchange Act: it is managed by a TSE board headed by the governor of the central bank, and no independent supervisory entity exists to oversee issuance and trading of securities. With few laws covering stock market operations, the TSE has itself introduced bylaws to cover insider trading, market manipulation, and disclosure and transparency requirements. However, these regulations are difficult to enforce without proper legislation. A draft law covering AML activities in the entire financial system, including trading in securities, awaits legislative approval.

The insurance regulatory framework is also outdated. Compulsory reinsurance, government control on tariffs, and the "specified proportions" approach to achieving prudent investments tend to result in excessive premiums and limit innovation and development of this predominantly state-owned industry.

Reform Agenda

The reform of the financial sector in Iran is incomplete, and further progress is needed to promote efficiency and facilitate the development of a dynamic and competitive financial sector that can meet the demands of an increasingly open and liberalized economic environment. Although reforms are needed in almost all segments of the financial system, the banking sector should be a high priority because of its size and role in allocating savings in the Iranian context.

Banking System

The reform agenda in the banking sector covers virtually all aspects of banking activity. Appropriate sequencing of reforms is important to their success in deepening financial intermediation. The highest priority is to complete the establishment of a risk-based supervisory framework; this should take precedence over steps in banking sector deregulation. Subsequently, state-owned banks will need to be restructured and their operational environment liberalized. Finally, state-owned banks might need to be recapitalized in connection with their possible privatization.

The reform of banking supervision has to be accelerated in connection with the ongoing liberalization and opening up of banking to private sector participation. This reform will help protect the banking system against the risks associated with the rapid credit growth that has occurred in recent years. The remaining reforms include the preparation, passage, and implementation of essential regulations pertaining to liquidity risk, asset classification, provisioning, and investments. The Banking Act should be amended to incorporate the concept of bank soundness among the objectives of bank supervision, to enlarge the range of sanctions to banks that do not comply with the regulations, to define banking and other services that banks and other financial entities are allowed to deliver, and to define the role of external bank auditors. Staff training and information technology development will be essential for successful implementation of these reforms. Finally, smaller deposit-taking institutions need to be brought under the supervision of the central bank as envisaged in the draft law on these institutions submitted to parliament.

Banking restructuring involves managerial, operational, and financial reorganization. Reforming the corporate governance of state-owned banks is key. As a first step, management of commercial banks should focus more on improving performance and strengthening the financial position of the banks. The undue influence by large public companies and bonyads on the management of banks has to be eliminated. Under the current system, managers of state-owned banks have little incentive or expertise to manage risks effectively; it will be important to provide adequate training in risk management, particularly in the area of credit risk.

Once risk-based banking supervision has been established and corporate governance of banks has been improved, further steps in deregulating the banking environment can be implemented. Rates of return on loans and deposits should be gradually liberalized, and the share of loans subject to sectoral allocation limits should be gradually reduced to zero. This will foster competition in the banking sector, improve pricing of risks, and contribute to more efficient allocation of financial resources. A reduction in administrative controls will also stimulate more effective use of existing Islamic finance instruments and development of new ones in line with recent international experience in this area (Sundararajan and Errico, 2002).

With respect to financial restructuring, the degree of undercapitalization of individual state-owned banks must be assessed on the basis of international norms, and the high lending concentration on large borrowers should be discouraged through strict implementation of the recently approved regulations on large exposures. Moreover, restructuring of banks will only be effective if accompanied by a restructuring of the large state-owned companies that are the major debtors. The program of restructuring and privatization of public companies will have to go hand in hand with banking system restructuring plans.

Capital Markets and Insurance

In light of the rapid increase in equity valuation, the reform agenda for capital markets needs to focus on tightening the supervision of securities issuance and on facilitating the market entry of properly supervised intermediaries. Efforts are under way to introduce a new capital market law that will cover securities issued inside and outside the TSE. This law intends to ensure the efficient functioning of securities markets and protect investors against unfair and fraudulent practices by requiring that adequate and timely information is provided to investors and the general public on companies that issue securities, and by regulating the activities of market intermediaries.

Another key reform should be to establish an independent securities and exchange commission. The development of market infrastructure (such as electronic trading, registration, and settlement of transactions) is also important and should go hand in hand with progress in regulatory oversight.

Similar to the other segments of the financial system, a risk-based insurance regulatory framework will need to be put firmly in place, and the Central Insurance Authority should divest itself of its reinsurance business and concentrate on regulation and supervision.

Capital Account Liberalization

The authorities have adopted a gradual approach toward capital account liberalization. The approach focuses mainly on attracting FDI, as reflected in the recent FDI law. Short-term flows, including portfolio investment, would be liberalized gradually. The draft portfolio investment law, which rightly takes a gradualist approach to liberalizing short-term inflows, is expected to authorize limited portfolio investment of nonresident institutional investors, with time limitations on the repatriation of principal capital.

Experience with capital account liberalization in developing countries indicates that the emphasis should be on reforms that help meet key preconditions for liberalization and enhance its benefits (Ishii and Habermeier, 2002; and Prasad and others, 2003). These preconditions include macroeconomic stability; an appropriate exchange rate regime; a strong and well-supervised financial system with developed and liquid capital markets; and significant improvements in key institutions, including legal framework and corporate governance.

Although Iran has made progress in all these areas, more work is needed to meet the preconditions for capital account liberalization. Macroeconomic stability must be firmly established—currently, inflationary pressures persist and the economy is vulnerable to sudden large changes in oil prices. Although a managed float exchange rate regime has been established, increased flexibility in the exchange rate is needed to deal with potential volatility in capital flows and fluctuations in oil revenue. Hedging instruments would need to be developed to increase the resilience of the financial system to exchange rate risk. As noted earlier, banks and the capital markets need to strengthen their capacity to monitor and assess the risks associated with volatile capital flows.

Conclusion

Although Iran has made progress in reforming its financial system, important challenges remain. The reform agenda for the period ahead should focus on restructuring the financial system and reducing its vulnerabilities. Reform of the banking system is of paramount importance; the highest priority is strengthening the supervisory framework and corporate governance of banks, which would ensure that a reduction in controls on credit allocation and rates of return will result in better financial intermediation. Managerial and organizational restructuring could be followed by recapitalization, privatization, and greater openness to foreign participation in domestic banks. Banking sector reform will need to be supported by the restructuring of the large state-owned companies that are the banks' major clients.

Proper supervision of the rapidly growing stock market is also needed, along with progress in developing the market infrastructure. The new capital market law is expected to address these issues and create a sound legal framework that will help foster the development of capital markets. Further capital account liberalization could be considered in step with progress in reducing inflation, reforming the financial sector, and introducing other supporting reforms.

CHAPTER

3 Moving toward a Market-Based Monetary Policy

Reform of the monetary policy framework in Iran is needed in connection with the efforts to liberalize the financial system (Chapter 2) and the authorities' objective of reducing the inflation rate to single digits. Drawing on academic research and the experience of other countries, as well as lessons learned from monetary policy management in Iran over the past 14 years, this chapter sketches a transition path from a system characterized by administrative controls and fiscal dominance toward one based on market incentives and signals. The chapter outlines a set of options for moving toward this objective by increasingly relying on indirect instruments of liquidity management consistent with Islamic finance principles.

The next section analyzes the main lessons from Iran's experience in conducting monetary policy and highlights the difficulties encountered by the Central Bank of Iran in achieving its monetary policy goals with limited instrument independence. The following section underscores the need to clarify monetary policy objectives and targets, enhance central bank instrument independence, improve the coordination between monetary and fiscal policies, and develop indirect instruments of liquidity management. The conclusion recommends a transition toward a monetary aggregate targeting.

Lessons for Monetary Policy Implementation

Institutional Setup

Current monetary policy formulation and implementation rely to a large extent on administrative controls in the context of fiscal dominance (Box 3). Although administrative controls are in part used to alleviate the inflationary impact of fiscal dominance and to favor the redistribution of credit resources according to government priorities, they are also motivated by the slow

Note: This chapter draws on the paper presented by Vitali Kramarenko and V. Sundararajan at a conference organized by the Islamic Development Bank in Tehran, February 16–21, 2004.

Box 3. Fiscal Dominance

Fiscal dominance has been an important source of high liquidity growth and inflation in Iran. Fiscal dominance stipulates that "monetary policy is subordinated to fiscal financing requirements" (Sargent and Wallace, 1981) or that fiscal policy is active whereas monetary policy is passive (Leeper, 1991).

In the context of Iran, an oil-producing country, the government budget relies to a large extent on oil export revenue earned in foreign currency, and until recently, domestic and foreign bond financing has been very limited. Two separate channels of fiscal dominance are at play in Iran: (1) central bank financing of government deficits ("pure" seigniorage) and (2) spending out of foreign-exchange-denominated oil revenue, which results in an increase in high-powered money.

- The role of pure seigniorage has been steadily declining. Direct central bank credit to the government was virtually discontinued in 1998/99 (except for bank recapitalization operations and some quasi-fiscal subsidies financed by the central bank), although central bank credit to nonfinancial public enterprises (NFPEs) continues on a small scale (Figure 6).
- Spending out of export oil revenue is equivalent to a foreign-financed expenditure and thus has a substantial effect on base money and real rates of return, especially under less-than-perfect capital mobility (Barnett and Ossowski, 2003). In Iran, the variations in expenditure and in its liquidity effects have been high (Figure 6). In this context, fiscal dominance is manifested in the inability of the central bank to offset sudden large changes in liquidity conditions stemming from fluctuations in government sales of foreign-currency-denominated oil revenue—mainly owing to their sheer magnitude. Insufficient development of appropriate instruments of liquidity control has also complicated liquidity management.

Figure 6. Sources of Base Money Growth, 1991/92–2002/03
(Increase as a percentage of the beginning-of-period base money)

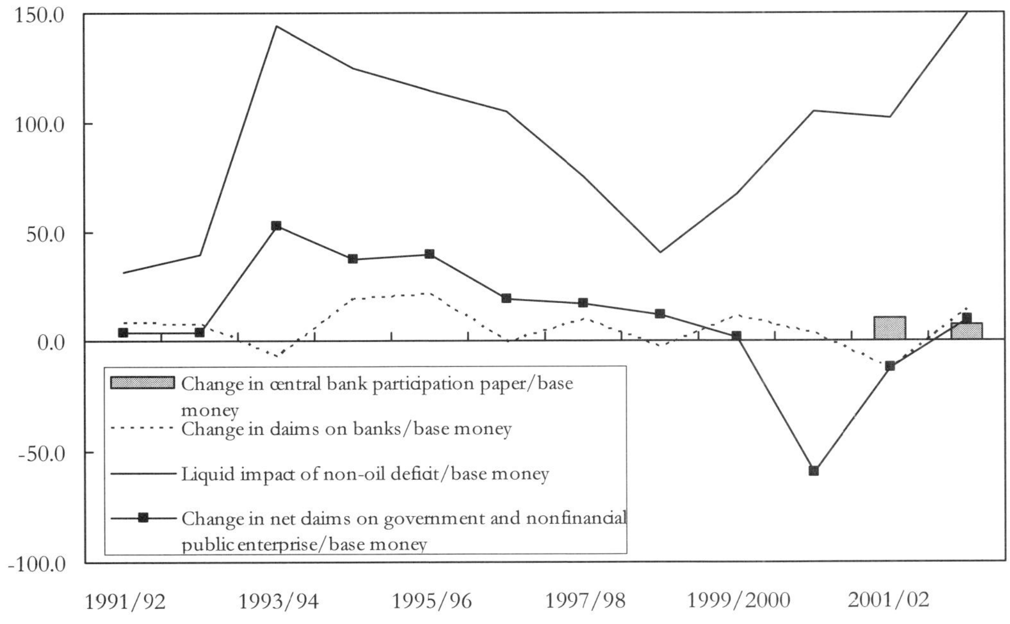

Sources: Iranian authorities; IMF staff estimates.
[1]Estimated by subtracting all sources of financing from oil revenue.

Table 12. Monetary Targets, 1989/1990–2004/05

(Percentage change)

	1989/90–1993/94		1995/96–1999/2000		2000/01–2004/05	
	FYDP I[1]	Outcome	FYDP II	Outcome	FYDP III	Outcome[2]
GDP[3]	8.1	7.4	5.1	3.3	6.0	5.6
M2	8.2	24.1	12.5	26.0	16.4	29.1
CPI	14.8	18.7	12.4	25.6	15.9	13.9

Sources: Central Bank of Iran; IMF staff estimates.

[1]Five-year development plan.

[2]Four-year averages, 2000/01–2003/04.

[3]At factor cost at constant 1997/98 prices.

progress toward developing money market instruments that are consistent with Islamic finance principles (i.e., Sharia).

The current approach to monetary policy formulation gives the government a decisive influence in setting specific monetary targets. In particular, FYDPs set annual targets for monetary growth and inflation that are approved by parliament and must be used by the central bank as benchmarks for formulating monetary programs. At the operational level, the Money and Credit Council (MCC) is responsible for key monetary policy decisions. Since 2005/06, the governor of the central bank has chaired the MCC; the minister of economy and finance is also a member, as are other ministers and government officials. Parliament and the government may issue directives for credit allocation, which could have implications for monetary policy implementation. In practice, the FYDP targets for broad money (M2) and inflation are usually revised by the MCC in its annual monetary guidelines. But even these revised targets are often inconsistent with fiscal financing requirements and other important decisions, in particular those on the administratively set rates of return and other direct controls on banking system activities.

Against this background, the central bank has not been able to meet its intermediate target for M2 since the inception of FYDPs. More important, these targets were exceeded by very large margins; as a result, the inflation rate objectives were not achieved during the first two FYDPs (Table 12).

Evolution of Monetary Policy Instruments

The role of direct instruments of monetary policy in Iran has gradually declined. Direct ceilings on refinancing facilities of commercial banks were abolished in 1991/92 and new directed, or "prearranged," credits of the central bank to commercial banks were de facto discontinued in 1998/99. The requirement for banks to hold government bonds was gradually relaxed during 1993/94, when banks were authorized to sell these bonds to the central bank.

Sectoral credit allocation limits and control on rates of return of state-owned banks have been gradually eased but remain significant. They are approved at the beginning of the fiscal year and almost never adjusted in the course of the year. The share of banking credit subject to sectoral allocation limits has been gradually reduced to 55 percent. Although state-owned banks may allocate 45 percent of loans without sectoral restrictions, they are still bound by administered rates of return that are fixed for each sector (Chapter 2).

Controlled rates of return on both loans and deposits were negative in real terms for most of the period under review, contributing to inflationary pressures and lack of progress in financial deepening (Figure 7). Real rates of return display a clear procyclical pattern—they are negatively correlated with the output gap—which means that monetary conditions become tighter during economic slowdowns and more relaxed during expansions (Figure 8). This is attributable to the limited flexibility of the rates of return combined with the liquidity impact of a procyclical fiscal policy.

Required reserves are high and differentiated by maturity of deposits, which has allowed the central bank to have better control of broader monetary indicators and to increase the demand for base money, thereby offsetting some of the impact of fiscal dominance. This situation in effect represents a tax on financial intermediation, because required reserves are only remunerated at 1 percent per year. The weighted average required reserve ratio declined to 16 percent in 2003/04 from 23 percent in 1990/91, which in part explains an upward trend of money multipliers.

Foreign exchange operations, excluding those with the central government, are an increasingly important element of monetary policy implementation. In 2002/03, the central bank's sales of foreign exchange in both domestic and offshore interbank markets amounted to about $13 billion (almost 100 percent of beginning-of-period base money).

The overdraft facilities of the central bank do not support monetary policy objectives. These facilities are in great demand by commercial banks because the interbank market is virtually nonexistent. In practice, the central bank accommodates liquidity shortfalls in the payment system without consistently

Figure 7. Real Rates of Return, 1991/92–2002/03
(In percent)

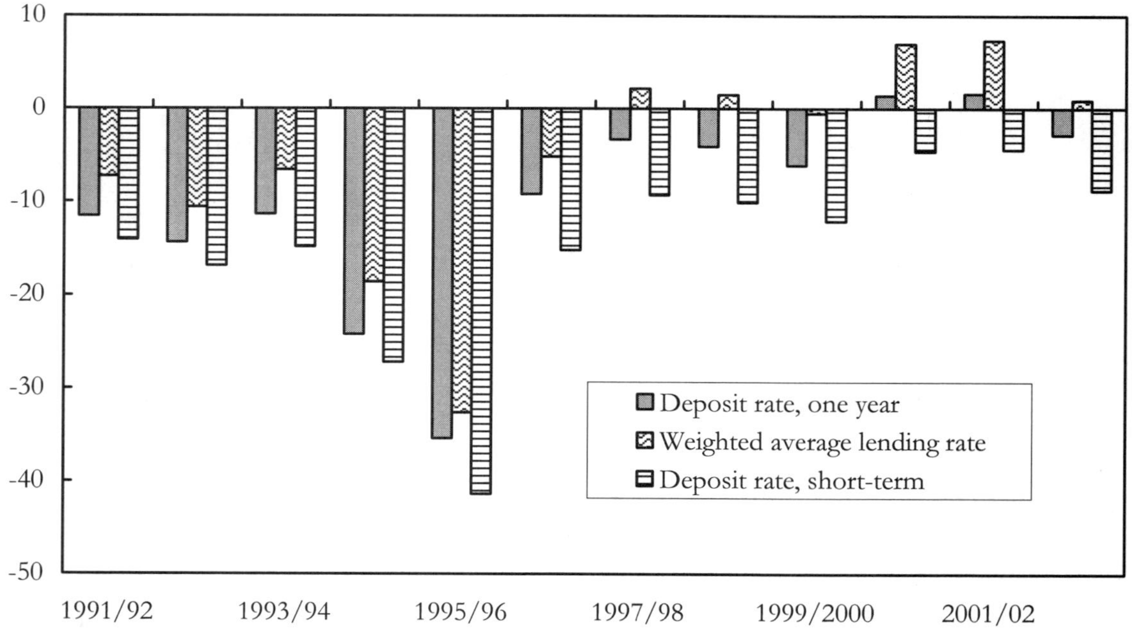

Sources: Iranian authorities; IMF staff estimates.

Figure 8. Non-Oil Output Gap and Real Rates of Return, 1991/92–2002/03
(In percent)

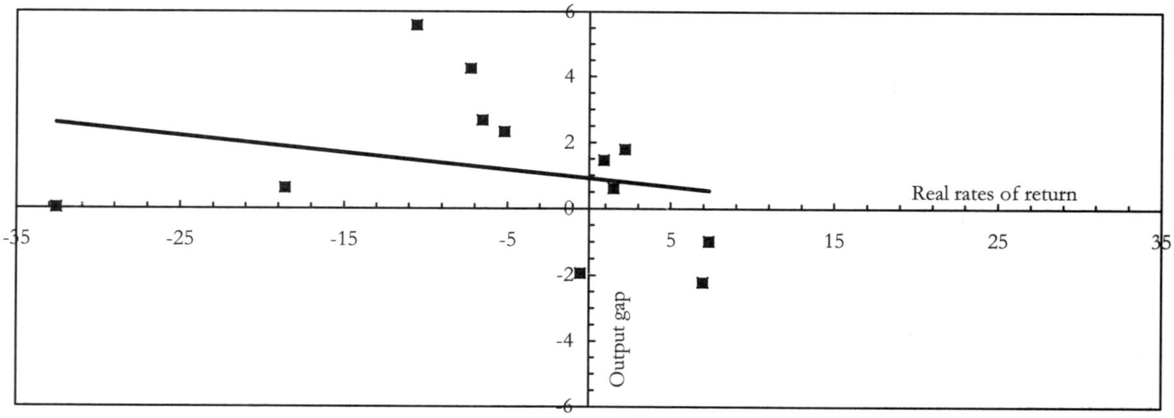

Sources: Iranian authorities; IMF staff estimates.

enforcing existing incentives against repeated large recourse to overdrafts. Until 1993/94, overdraft rates were set at only 2 percentage points above the directed credits; in that year, they were replaced by a progressive schedule of overdraft rates at 20, 24, and 30 percent, depending on access levels. If overdraft periods exceed three days, an additional 4 percentage points are added to each tier. Despite revisions to other administered rates in the banking system and sharp fluctuations in annual inflation since 1993/94, the overdraft rates have not been revised. Moreover, on some occasions, overdraft penalty payments have been waived. Frequent recourse to relatively cheap or even penalty-free overdraft facilities often puts the central bank in an accommodating position. There is also a standing credit facility providing financing up to one year.

The standing deposit facility (open deposit accounts) introduced in 1998/99 has played a marginal role in central bank operations. This facility has been used to regulate liquidity fluctuations within the year; but by year's end, when most banks experienced liquidity pressures, the deposits were usually drawn down substantially. In 2003/04, the central bank attempted to auction deposit facilities to commercial banks, but there was no demand for such instruments, in part because of the unattractive terms.

Central bank participation papers (CBPPs)—which are compatible with Islamic finance principles—were first issued in March 2001 (Table 13). The bearer securities are issued in parcels of Rls 1, 2, 5, and 10 million, which reflects their retail focus. Maturities are 6 or 12 months, with nontaxable quarterly coupon payments. The CBPPs are issued based on a portfolio of completed infrastructure projects previously financed by the central bank credit to government; they yield a predetermined rate of return presumed to approximate the returns on underlying assets. This means that the CBPPs are backed by underlying central bank claims on the government. The objective was to create a marketable money market instrument that would empower the central bank to regulate liquidity and provide a viable instrument for liquidity management by commercial banks, leading to the development of an interbank market (Ul Haque and Mirakhor, 1998). However, the final design turned out to be somewhat different.

CBPPs have been issued only to nonbanks in the primary market, at preannounced rates of return. Commercial banks are obliged to rediscount them in the secondary market at par and to guarantee the initial yield to maturity. In other words, CBPPs cannot be traded in the secondary market at prices different from par and thus represent a highly liquid instrument held by nonbanks. Although to some extent these instruments have helped absorb liquidity created by government sales of foreign-currency-denominated oil revenue, their high cost to the central bank raises questions about the sustainability of their growing stock. Moreover, at times, the central bank has

Table 13. Government Participation Papers and Central Bank Participation Papers, 1997/98–2002/03
(In billions of rials, unless otherwise indicated)

	1997/98	1998/99	1999/2000	2000/01	2001/02	2002/03
Government Participation Papers (national)						
Amount issued (gross)	2,174	2,500	1,927	0	0	2,400
Average maturity (years)	3.0	3.0	4.0	. . .	. . .	5.0
Average rate of return (in percent, per year)	20.0	20.0	19.0	. . .	. . .	15.0
Stock (end-of-period)	2,174	4,674	6,601	4,677	4,427	6,827
Central Bank Participation Papers						
Amount issued	0	0	0	0	12,359	17,052
Average maturity (years)	. . .	. . .	. . .	. . .	0.8	1.0
Average rate of return (in percent, per year)	. . .	. . .	. . .	. . .	17.5	17.0
Stock (end-of-period)	. . .	. . .	. . .	. . .	9,443	17,052
In percent of beginning-of-period base money	. . .	. . .	. . .	. . .	10.5	17.6

Source: Central Bank of Iran.

not been able to achieve the targeted amount of issues at the rate of return fixed by the MCC. These design features of CBPPs have limited their effectiveness.

Government participation papers (GPPs) were first issued in 1998 (Table 13), but they did not play a major role in attenuating the impact of fiscal dominance or stimulating the development of money markets. (Participation papers of municipalities and various ministries and public enterprises have been authorized since 1994.) A GPP is an instrument used to finance nonspecific government infrastructure projects by providing investors a temporary (equal to the maturity of the paper) equity stake in the underlying assets. The government promises to pay on maturity a return that approximates the rate of return on the underlying asset, which should be at least equal to the private sector rate. GPP is a different instrument than the government bonds issued in the 1980s. GPPs are not designed for use by the central bank to manage liquidity but were primarily intended to finance central government infrastructure projects. GPPs are issued in the primary market at preannounced fixed rates of return with a five-year maturity, and the outstanding stock is modest. Banks are obliged to rediscount GPPs in the same manner as CBPPs (see above). The tax-adjusted rates of return on GPPs are below those on CBPPs, despite the much longer maturity of the former. This implies a negatively sloped yield curve of rates of return, which has reduced the attractiveness of GPPs in the presence of high uncertainties over future inflation developments.

Monetary Policy since 2002

The 2002 exchange rate unification and the establishment of a managed float exchange rate regime raised the issues of the appropriate nominal anchor and related supporting policies. Although the central bank has increasingly focused monetary policy implementation on M2, this policy has not yet achieved sufficient credibility to effectively anchor inflationary expectations. In the aftermath of the 2002 exchange rate unification, exchange rate considerations continued to dominate, initially out of concern for the stability of the nominal rate and then to preserve competitiveness through a gradual nominal effective depreciation of the exchange rate to compensate for past inflation differentials. However, this dual-objective policy became difficult to sustain in the face of an increased supply of foreign exchange stemming from fiscal relaxation and FDI inflows. In an attempt to contain the growth of monetary aggregates while continuing with nominal exchange rate depreciation, the central bank began to use CBPPs to mop up excess liquidity at relatively attractive fixed rates of return, thereby directly bearing the cost of sterilization operations. With the rapid increase in the CBPP stock, these operations became costly and less effective at offsetting large injections of oil revenue into the system. As a result, the amount of unsterilized purchases became a function of exchange rate objectives, and the control over monetary aggregates weakened. The policy of nominal depreciation did not prevent the REER from appreciating by 7.5 percent during 2002/03–2003/04,[1] reflecting a large inflation differential between Iran and its trading partners.

During 2003/04, the policy on rates of return conflicted with monetary policy targets. Negative real rates of return on loans for the major sectors of the economy—such as industry, mining, and agriculture—helped fuel credit demand growth, which the central bank accommodated through its overdraft facilities. It is clear that the rapid credit growth, together with large unsterilized purchases of foreign exchange from the government, also contributed to nominal exchange rate depreciation.

Next Steps

Monetary Policy Framework

There is a pressing need to improve the monetary policy framework and to enhance coordination between fiscal and monetary policies. Drawing on the extensive academic literature and operational experience of central banks in

[1]The REER was compiled using a trade-weighted index that excludes oil trade.

market-based financial systems (Walsh, 2003), this subsection highlights the general principles of an effective monetary policy framework; the next subsection elaborates on the implications of using Sharia-compliant monetary policy instruments.

Monetary Policy Objectives

Since Iran has chosen a managed float exchange rate regime,[2] price stability should be an overriding objective of monetary policy. Possible intermediate targets consistent with this objective include a monetary aggregate or a measure of consumer price inflation. Targeting a monetary aggregate such as broad (M2) or narrow (M1) money is perhaps more familiar to policymakers in Iran and has less stringent institutional and policy requirements than targeting, say, an annual percentage change in the CPI.[3] The M2 intermediate target could be set jointly by the government and the central bank, consistent with the desired outcomes for inflation. Greater simplicity in the transparency and accountability requirements under monetary targeting is a distinct advantage. In contrast to inflation targeting (Schaechter, Stone, and Zelmer, 2000), adherence to monetary targets is easier to monitor. Timely publication of data on monetary aggregates is an important accountability requirement, but it is easier to prepare, interpret, and understand than subtle explanations of monetary policy actions needed to achieve inflation objectives in the presence of complex transmission mechanisms.

A transition toward monetary aggregate targeting could be considered notwithstanding money velocity instability in the recent past. The empirical study of Celasun and Goswami (2002) finds that inflation is affected by real money and output growth but also depends on exchange rate developments and the degree of deviation of real money demand from its equilibrium level. The implication of these findings is that inflation forecasts are subject to uncertainty, and it would be difficult for the central bank to meet an inflation target by adhering strictly to intermediate M1 or M2 targets. This does not mean, however, that M1 or M2 targeting should not be considered an option. An indicative inflation objective could be formulated as a band in order to accommodate forecast errors in setting an annual target for M1 or M2 growth.

---

[2]See Sundararajan, Lazare, and Williams (1999) and Celasun (2003) for a justification of this choice.

[3]A discussion of the prerequisites for establishing an inflation-targeting framework and the assessment of Iran's compliance with these prerequisites is beyond the scope of this paper. See Jbili and Kramarenko (2003) for a detailed analysis.

The central bank needs to determine an operational target that it can directly control, such as base money or money market rates of return. The latter, however, are not readily available because of the slow development of money markets in Iran. Thus, base money could be initially selected as an operational target. Given the instability of money multipliers,[4] the base money target should be revised periodically in light of new information to maximize the chances of hitting the intermediate target for M1 or M2. It is also important to ensure that fiscal policy is consistent with the need to achieve the base money operational target, which would require a careful assessment of the liquidity impact of fiscal operations at the budget preparation stage. Specifically, the authorities would need to ensure that the size and composition of the non-oil deficit financing is consistent with the operational and intermediate targets of monetary policy.

Instrument Independence of the Central Bank

As mentioned earlier, current legislation provides the central bank with limited authority to use monetary policy instruments—such as rates of return or the amount of CBPP issuance—without prior approval from the MCC. This institutional arrangement does not give the central bank the needed flexibility to deal rapidly with changes in money and credit conditions during the year. Moreover, the fact that the government or parliament can issue credit directives undermines the ability of the central bank to meet its intermediate targets. Granting the central bank instrument independence—defined as day-to-day independence in using all relevant instruments needed to achieve intermediate targets, subject to a possible override provision for the government—is essential for successful implementation of monetary aggregate targeting. Such independence will need to be combined with stringent accountability requirements to various layers of authority and the public. There is also a need to establish transparent procedures for resolving potential conflicts between monetary policy and broader economic policy objectives.

The range of instruments at the disposal of the central bank should be broadened, and the existing instruments should be adapted to the new framework. The required reserve ratios should be unified and reduced to lower the cost of financial intermediation, provided that offsetting measures are put in place to mop up excess liquidity. Access to the standing credit facilities should be tightened and made more onerous to discourage frequent

---

[4]The multiplier instability has been caused recently by a significant increase in the use of prepaid checks for payments, which reduced the demand for banknotes. Although checks are not to be endorsed more than once, in practice they are endorsed many times before they are deposited in a bank.

use. Indirect instruments of monetary policy need to be redesigned to gradually become the preferred instruments in order to facilitate the emergence of a benchmark rate of return (the following subsection includes a description of options that could be implemented in the short run). Once money markets grow in depth and experience, the operational target could be changed from base money to the rate of return on an appropriate money market instrument. Foreign exchange operations will continue to be important, but the central bank needs to gradually shift the emphasis in these operations from the exchange rate to base money by more actively using indirect instruments of monetary policy and tolerating greater fluctuations in the exchange rate.[5] Monetary instruments alone, however, are unlikely to be sufficient to sterilize the liquidity impact of injections of government oil revenue into the system (see Box 3) or of large capital inflows; thus, fiscal policy actions would also be needed. In the long run, incorporating market-based mechanisms into the design of government participation papers and increasing their outstanding volumes and liquidity would also facilitate the deepening of financial markets.

Sharia-Compliant Indirect Instruments of Monetary Policy

This subsection reviews options for developing Islamic money market instruments in Iran. A number of difficulties arise in designing short-term financing instruments that are Sharia-compliant (that is, are interest-free, rely on profit and loss sharing linked to real transactions, or are based on purchase and resale contracts) and whose value can be determined at a high frequency to facilitate short-term trading and money market operations. Several central banks—notably in Malaysia, Sudan, and Bahrain—have developed Islamic financial instruments to facilitate liquidity management (by the central bank as well as by commercial banks) and public borrowing (Majid, 2003). At the same time, there has been a growing use of asset securitization techniques to design Islamic securities for issuance in regional and international capital markets (Hassan, 2002). This has opened the door for developing short-term instruments for monetary operations.

Effective market-based monetary operations require an instrument with the following characteristics:

- The instrument must be relatively risk-free. It must be able to serve as a benchmark to price more risky instruments of varying maturities and strongly influence the marginal cost of funds for banks.

[5]The medium-term issues of competitiveness and the appropriate level of the exchange rate are beyond the scope of this chapter, which is focused on short-term aspects of monetary policy implementation.

- A sufficient supply of the instrument must be available to meet both monetary policy needs and the portfolio needs of investors.

- The instrument must be widely held by both banks and nonbanks to support a liquid market.

- The payment settlement system must be robust and reliable to facilitate trading in the instrument.

On the basis of these criteria, a structure that securitizes a range of Islamic financial contracts seems the most promising for monetary operations; all other market instruments do not meet one or more of the criteria for effective monetary operations.

Purely equity-based instruments (*Musharaka*), or structures containing only cash-flow rights from government ownership in enterprises can carry high returns that raise costs to the government. Also, the volume of issuance of such instruments may not be sufficient for monetary policy purposes, insofar as the issue amounts are limited by the extent of government ownership in high-quality enterprises.

Purely commodity-resale-type instruments (*Murabaha*) and participation papers with guaranteed minimum returns may not trade at prices different from par under Islamic finance principles and thus cannot be a reliable basis for developing interbank money markets.

Purely debt-type contracts (*Mudarabah*), such as interbank deposit placements linked to bank profits, are not suited for the liquidity-absorption operations of central banks owing to the difficulties of linking returns to central bank profits. Moreover, differences in perceived bank risks might limit the volume of interbank placements.

Thus, securities based on a mixture of contracts representing equity- (Musharaka), debt- (Mudarabah), and leasing-type financing (*Ijara*) have the best chance of being issued in sufficient volume, achieving adequate market liquidity, and providing sufficient flexibility in the mix of risks and return. The mix of contracts should be transparent to allow investors to assess risks and form expectations of returns based on expected performance of the underlying cash flow.[6]

---

[6]Also, the proportion of pure equity-type contracts in the overall asset mix could be higher for longer maturities, thereby varying the risk-return trade-off by maturity.

The ongoing issues of CBPPs in Iran can be transformed over time to become an effective instrument of monetary management by

- identifying a wider range of government assets and cash flows that can be securitized,
- strengthening the coordination of public expenditure management and the government financing program to ensure an optimal combination of assets that can be securitized,
- adopting a high-quality and transparent accounting and disclosure framework for communicating the value and returns on the underlying assets,
- adopting auction-based primary issuance that reflects market expectations in the price of the security,
- supporting the liquidity of the instrument in the secondary market through repurchase facilities, and
- organizing efficient trading and payment settlement arrangements.

Such newly designed Sharia-compliant instruments would overcome the current constraints on Iran's CBPPs and allow more flexible rates of return and better-functioning secondary markets to emerge, thereby facilitating more effective monetary and public debt management.

Conclusion

The current system of monetary policy formulation and implementation still largely relies on administrative controls in a context of fiscal dominance. Although administrative controls are used to alleviate the inflationary impact of fiscal dominance and to direct credit resources according to government priorities, they are also motivated by the slow progress in developing money market instruments consistent with Islamic finance principles.

The need for a properly sequenced financial liberalization and the stated objective of reducing inflation call for major changes in the monetary policy framework in Iran. Initial steps in this area could include the development of monetary aggregate targeting. Central bank instrument independence, stringent accountability requirements, and the development of indirect instruments of monetary policy are the major ingredients of success in this area. Although developing liquid money market instruments consistent with Islamic finance principles may be a difficult undertaking, the obstacles are surmountable, as evidenced by the experience of other countries.

Reforming monetary policy alone will not remove inflationary pressures or enhance financial intermediation. The elimination of fiscal dominance; the restructuring of the banking system, with a greater emphasis on private sector participation and competition; and other institutional reforms are key to achieving sustainable low-inflation growth.

CHAPTER

4 Approaches to Assessing the Exchange Rate Level

Following a long period of multiple and unstable exchange rates in Iran, the exchange rate was successfully unified in March 2002. Since then, Iran has adopted a managed float exchange rate regime and eliminated most exchange restrictions for current account transactions, culminating with the acceptance in August 2004 of obligations under Article VIII, Sections 2, 3, and 4 of the IMF's Articles of Agreement. Like many other developing countries with managed floats (Hinkle and Monteil, 1999; and Ishii, 2003), the Iranian authorities continue to pay close attention to the nominal exchange rate. This raises theoretical and practical questions about how to determine the appropriate level of the exchange rate and how to initiate corrections if misalignments are detected.

Finding indicators to assess the exchange rate level, let alone estimating an equilibrium REER, is a major challenge in the Iranian context. The difficulties stem from the large share of oil exports in trade and the various distortions created by government intervention in economic activity. Theoretical and empirical cross-country research provides guidance on how to integrate the terms of trade in the analysis of exchange rate level in commodity-producing countries, but these studies are less relevant for a country like Iran, because of the prevalence of price and interest rate controls and the government's direct intervention in allocating resources to priority sectors and enterprises. Frequent changes to trade policy and exchange restrictions, as well as numerous tax exemptions and subsidies, add another layer of complexity.

This chapter suggests a range of indicators that could be used to assess the exchange rate level. However, because of the difficulties noted above, no single indicator is given prominence; the particular combination of these indicators that would be best for assessing the appropriate level of the exchange rate remains a matter of judgment.

Note: Prepared by Vitali Kramarenko.

Figure 9. Imports, Market Exchange Rate, and Trade and Payment Restrictions, 1992/93–2003/04

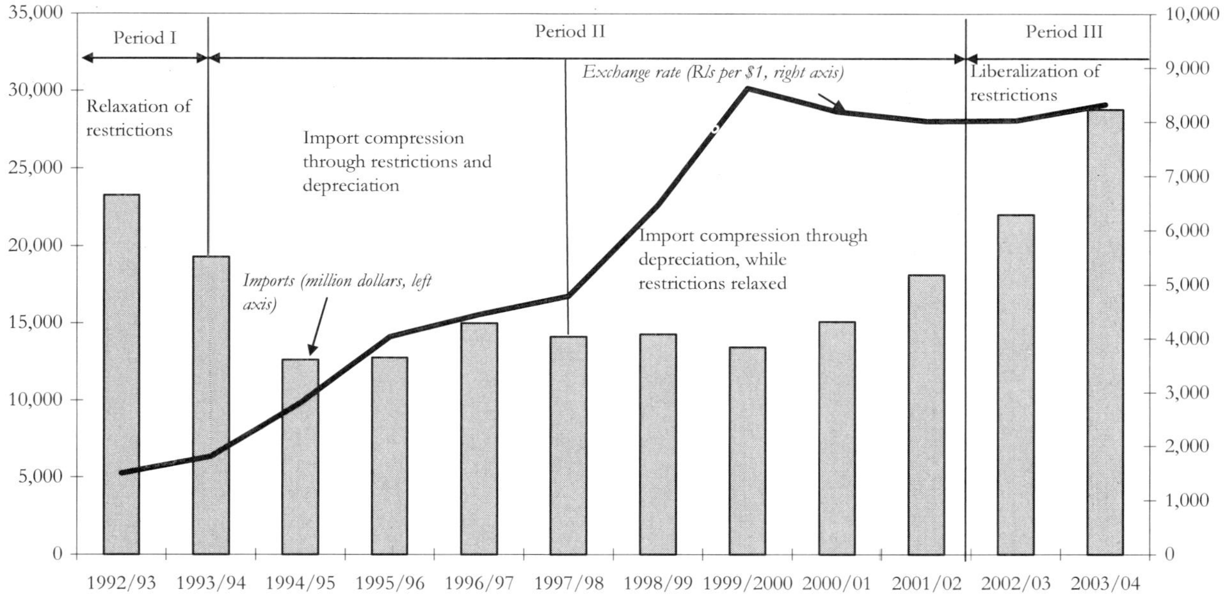

Sources: Iranian authorities; IMF staff estimates.

The chapter begins with a brief review of the institutional setup of the foreign exchange market during 1993–2003, which provides background and identifies key stages in the evolution of the exchange rate regime. Econometric techniques are used to analyze developments in the REER, including testing a purchasing power parity hypothesis and, in a more refined framework, incorporating productivity trends, terms of trade developments, and monetary policy variables. This empirical investigation is complemented by a discussion of other possible means of assessing the exchange rate level, including growth performance of the tradables sector and comparison of the actual external current account balance against a "norm" value derived from cross-country studies. The chapter concludes by suggesting that the level of the exchange rate in 2003/04 does not appear to have been misaligned, judging from the good performance of the tradables sector and the relative proximity of the current account balance to its long-term "normative" value.

Exchange Rate Policy in Iran since 1993

Since 1993, the exchange rate regime in Iran has undergone numerous changes in three distinct periods (Figure 9; Sundararajan, Lazare, and Williams, 1999; and Celasun, 2003).

The first period coincides with the first attempt at exchange rate unification, which lasted for nine months during March–December 1993. This effort failed mainly because of an expansionary fiscal policy and an unfavorable external environment. Following the exchange rate unification, large foreign exchange losses associated with subsidized imports of basic consumption commodities and payments on public and publicly guaranteed debt resulted in sizable quasi-fiscal losses for the central bank. The monetization of these losses led to an acceleration in money growth to 35 percent in 1993/94 from 25 percent in 1992/93. This, along with some limited capital account and trade liberalization measures, put pressure on the foreign exchange market. The central bank faced difficulties in meeting the growing demand for foreign exchange without a large exchange rate adjustment,[1] as oil prices started to decline and access to trade credits was significantly curtailed.[2] The disequilibrium in the foreign exchange market culminated in a freeze on the official exchange rate that was applied to foreign exchange purchases used for essential imports and debt service payments. At the same time, an officially recognized market-determined exchange rate—the dealers' market rate—emerged (Figure 10).

During the second period (1994–March 2002), various combinations of multiple exchange rates existed (Table 14). Nontariff import restrictions were gradually tightened through 1997, including strict import licensing and positive import lists for each segment of the foreign exchange market. Also, export surrender and repatriation requirements were introduced. Despite tighter trade and payment restrictions, the parallel market nominal exchange rate continued to depreciate sharply, as macroeconomic policies remained highly inflationary.

After 1997, a number of trade restrictions were gradually relaxed, which further fueled parallel market exchange rate depreciation. By 1998/99, macroeconomic policies were tightened and the parallel market exchange rate, as well as an increasingly market-determined exchange rate in the Tehran Stock Exchange, stabilized.[3] In 1999, all administrative controls on the TSE exchange rate were lifted, laying the basis for a smooth transition toward a unified exchange rate system. Despite these liberalization efforts, stringent licensing requirements and a positive list for imports remained in place until March 2002. The parallel market for foreign exchange remained

[1]Gross official reserves fell below $3 billion, or 1.4 months of next year imports, at end-March 1994.

[2]Net inflows under letters of credit of $2.5 billion in 1992/93 were followed by net outflows of $5 billion in 1993/94.

[3]Beginning in 1997, a growing number of foreign exchange transactions were allowed to take place at the TSE exchange rate. Foreign exchange was not traded directly. Rather, certificates of export proceeds, which could be used for authorized imports, were allowed to be traded in the TSE at increasingly market-determined exchange rates.

Figure 10. Exchange Rate Developments, 1993–2002
(Rials per U.S. dollar)

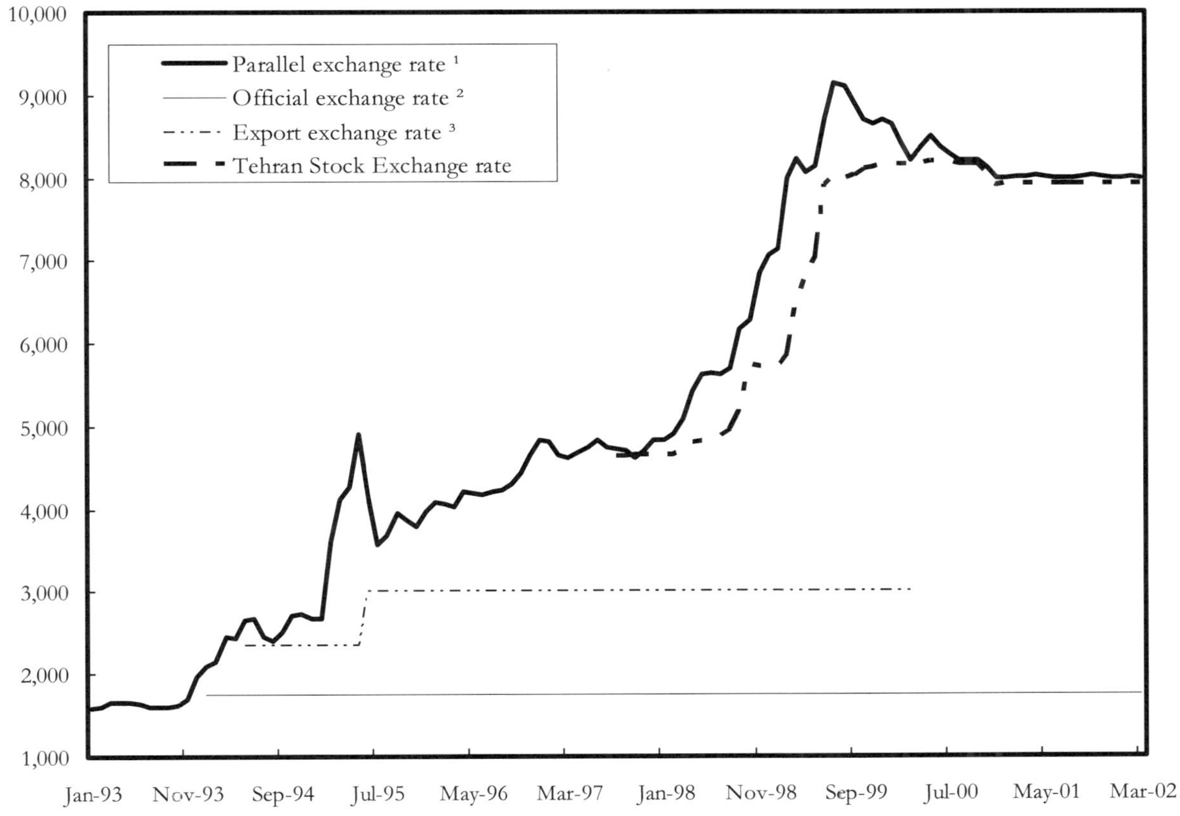

Sources: IMF, Information Notice System; Iranian authorities.
[1]Initial observation: January 1993.
[2]Initial observation: December 1993; final observation: March 2002.
[3]Initial observation: May 1994; final observation: March 2001.

active up to the March 2002 unification inside the country and offshore, even though the premium over the TSE rate was on a declining trend.

The third period started with the March 2002 exchange rate unification and a substantial reduction in import restrictions. During 2003, some capital account liberalization measures were undertaken as well (Chapter 2). The authorities opted for a managed float exchange rate regime, taking into account the nature and magnitude of the shocks the economy was likely to face, as well as the structural characteristics of its goods, labor, and financial markets. Since Iran has been subject to both real and nominal shocks in the context of significant nominal rigidities, a managed float exchange rate

Table 14. Chronology of Exchange Rate Developments, 1991–2003

Starting Date	Exchange Rates	Rials per U.S. dollar	Transactions Taking Place at the Rates
1/21/91	Basic official rate	70	Oil exports, imports of essential goods, imports for priority projects, official debt service, and government-supported students.
	Competitive rate	600	Imports of intermediate and capital goods not included above as well as related services.
	Floating rate	Floating—determined by banks based on free market rate	Non-oil exports, imports not covered by the above official rates, and most service payments and receipts.
	Free nonbank market rate	Market-determined	Foreign exchange freely bought and sold.
3/21/93	Floating rate (unified official rate)	Managed float	All transactions (previously contracted debt were serviced at pre-March 1993 exchange rates and imports of certain essential goods were effected at the previous basic official rate through a government subsidy account during 1993/94).
	Free nonbank market rate	Market-determined	Foreign exchange freely bought and sold.
12/21/93	Official rate	1,750	All transactions initially. Gradually, the eligible payments became limited to imports of essential goods; payments for essential services, and official debt service.
	Free nonbank market rate	Floating	All other transactions, including non-oil exports, tourist receipts, payment for nonessential goods and services, and transfers.
5/4/94	Official rate	1,750	Oil exports, imports of essential goods, payments for essential services, and official debt service.
	Official export rate	Initially Rls 50 below authorized dealers' rate, soon fixed at Rls 2,345.	Non-oil exports and imports on a positive list.
	Authorized dealers' market rate	Floating (Rls 2,680 on 12/31/94; as depreciated as Rls 6,200 in 5/95).	All other transactions, including non-oil exports, tourist receipts, payment for nonessential goods and services and transfers.
	Parallel offshore market rate	Market-determined (Rls 3,000–3,100 on 12/31/94).	Foreign exchange freely bought and sold.
5/20/95	Official rate	1,750	Oil exports, imports of essential goods, imports for priority projects, payments for essential services, and debt service on priority projects.
	Official export rate	3,000	All other transactions, including non-oil exports, service receipts, and imports and service payments not covered by the official rate.
7/11/97	Official rate	1,750	Oil exports, imports of essential goods, imports for priority projects, payments for essential services, and debt service on priority projects.
	Official export rate	3,000	All other transactions including non-oil exports, service receipts; and imports and service payments not covered by the official rate.

Table 14 *(concluded)*

Starting Date	Exchange Rates	Rials per U.S. dollar	Transactions Taking Place at the Rates
7/11/97	Exchange rate for import certiticates	4,600	Import certificates have been allowed to be traded through the TSE—up to 100 percent for carpet exporters and up to 50 percent for other non-oil exporters.
1/98	Conversion of export proceeds	Determined at the TSE	All export proceeds could be sold at the TSE.
1/99	Tehran Stock Exchange (TSE)	Market-determined	The TSE rate is allowed to depreciate and the queuing in the TSE is abolished.
3/00	Export exchange rate	3,000	Abolished.
3/02	Exchange rate unification	Market-determined	A unified interbank market established.
2003	Unified rate	Market-determined	Opening-up of the capital account for some ouflows through the off-shore foreign exchange market.

Source: Based on information provided by the Central Bank of Iran.

regime appeared appropriate, because it was likely to reduce the output costs of adjustment to shocks.[4] A low degree of dollarization of the financial system (Chapter 2) was another important consideration for adopting an exchange rate regime with some degree of flexibility.

The 2002 unification has been successful, as evidenced by the stability of the nominal exchange rate and the significant decline of the premium in the parallel exchange market. The success is mainly due to the following factors. First, Iran initiated the exchange regime change from a position of strength, benefiting from high oil prices, renewed reform efforts, and a few years of experience with a market-determined exchange rate in the TSE market, which had operated in the context of a dual exchange rate system. Second, the Iranian authorities were able to minimize the pass-through to prices of the depreciation of the official exchange rate, relying on resources previously accumulated in the Oil Stabilization Fund (OSF) and high oil revenue to subsidize basic consumer goods. As a result, following the exchange rate unification and the move to a managed float, the exchange rate quoted in the interbank market remained relatively stable, which helped bolster business confidence and ensure a smooth transition to the new system.

After the fall of 2002, the authorities initiated a gradual depreciation of the nominal effective exchange rate vis-à-vis a basket of currencies, while

[4]See Celasun (2003) and Jbili and Kramarenko (2003) for detailed analyses and other factors in the choice of exchange rate regime in Iran.

reverting to an expansionary policy mix. At the same time, a significant reduction of nontariff import barriers and some liberalization of capital outflows (Chapter 2) helped, to some extent, contain pressure on the nominal exchange rate to appreciate against the background of increasing government sales of oil-related foreign exchange to the market to boost expenditure. Despite these measures, double-digit inflation persisted.

REER-Based Measures of Competitiveness

The analysis in this chapter uses exchange rates from the IMF's Information Notice System, which splices data on various representative market-related exchange rates and other exchange rate series provided by the central bank.[5] The overly appreciated official exchange rate fixed at Rls 1,750 per US$1 during 1993/94–2001/02 is not used, because it was unrelated to fundamentals and mainly served redistribution purposes, which, in practice, could have been achieved through explicit taxes and subsidies (Mullin, 1993). The official rate mainly applied to energy-related exports and subsidized imports (whose pricing and distribution were subject to tight government oversight) and to debt service payments on public and publicly guaranteed debt. To calculate the REER, CPI inflation differentials with trading partner countries are used. Weights are derived from the composition of trade (REER1). To check the robustness of results to changes in assumptions on trade weights, an REER using weights from the non-oil trade composition is also calculated (REER2). Excluding oil trade from the REER calculations reduces the relative weights of the euro area and Japan—the major destinations of oil exports—and provides a better indicator of competitiveness of the non-oil economy in Iran.

Relative Price Indicators of Exchange Rate Level

The REER and the PPP

Purchasing power parity is a possible yardstick for exchange rate assessment. It is based on the law of one price (Dornbusch, 1987), which, under the absolute PPP definition, stipulates that every good should have the same price once converted into a common currency unit. If the law of one price holds, the values of identical baskets of goods should be the same across countries once they are converted into a common currency unit.

---

[5]The nominal effective exchange rate (NEER) for Iran was calculated using data on the unified exchange rate during 1993, the free market dealers' rate (1994–mid-1995), and, after the closure of the dealers' market, the parallel market exchange rate (mid-1995–97), the TSE rate during 1998–2002, and the unified interbank rates thereafter.

The most widely used methodology to confirm or reject the PPP is based on the analysis of the time-series properties of the REER, which is assumed to measure changes in price level differences between a country and its trading partners (Rogoff, 1995). If the REER series is stationary (that is, does not have a stochastic trend), the PPP should hold. Even if the REER series is not stationary, the speed of convergence of the REER toward its equilibrium should be fast enough (say, below one year) for the PPP to hold. Slow convergence would imply that disequilibrium could persist in the medium and long term (Rogoff, 1995). This version of tests is consistent with a *relative* PPP formulation, which accounts for changes in the CPI rather than changes in the prices for specific goods and services.

We apply this methodology to Iran's REER series. Both REER series show a clear cyclical behavior with an upward trend (Figure 11), and the visual observations are confirmed by statistical tests. Both are nonstationary I(1) processes based on Augmented Dickey-Fuller (ADF) test statistics.[6] As a result, a no-PPP hypothesis cannot be rejected in the case of Iran. This evidence is similar to the findings for other commodity producers (Cashin, Céspedes, and Sahay, 2002; and Chen and Rogoff, 2002).

It may be that these tests are not powerful enough to reject the null hypothesis of nonstationarity, and the series could actually be stationary (Kwiatkowski and others, 1992). If one supposes that the REER is stationary, given the short length of the series, the next step would be to determine the speed of REER adjustment to a unit shock. For this purpose, an AR(1) ordinary least squares (OLS) regression was run to estimate the half-life of a unit shock or impulse (HLS). The estimated autoregressive coefficient α is equal to 0.98 for both series; thus, the HLS[7] was equal to 35 months. This length is similar to those of most commodity-based currencies (Cashin, Céspedes, and Sahay, 2002). As mentioned, a three-year half-life for adjustment is difficult to reconcile with the traditional PPP formulation, which allows for only short-term deviations.

The results of the tests are not consistent with the PPP theory conjecturing that the REER should be a mean-reverting I(0) process with a relatively rapid pace of adjustment. Although the PPP theory is rejected for Iran, assessing the REER level with respect to its long-term mean still provides a point of departure for further analysis. By March 2004, REER1 and REER2 were above their long-term means by 22 and 29 percent, respectively. The next step is to assess whether such a real appreciation is justified on the basis of additional considerations.

---

[6]ADF statistics are –1.24 (0.65) and –1.23 (0.66) for REER1 and REER2 in logs, respectively, with *p*-values in brackets.

[7]HLS = abs (log ½ /log (α)).

Figure 11. Real Effective Exchange Rate Indices, 1993–2004

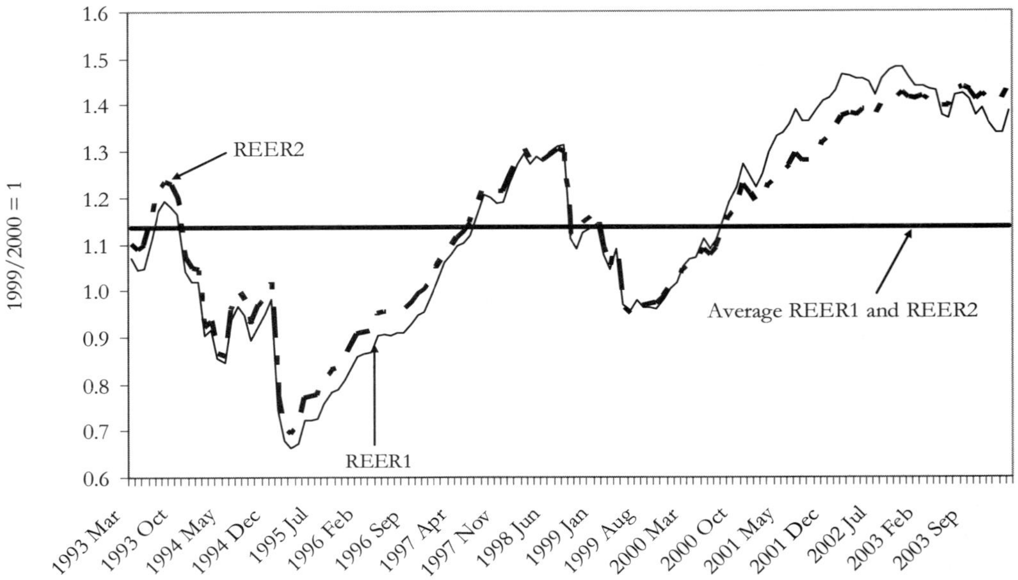

Sources: International Financial Statistics; IMF staff estimates.

The REER and Real Long-Term Fundamentals

There is a strand of theoretical literature that explains drifting REER in developing countries by productivity differentials[8] and terms of trade (De Gregorio and Wolf, 1994; and Obstfeld and Rogoff, 1996).[9] Drawing on such theoretical models, empirical literature usually looks for a long-term relationship among REER, productivity differentials, and terms of trade. Based on co-integration tests (see Rogoff, 1995; Chen and Rogoff, 2002; and Cashin, Céspedes, and Sahay, 2002), there is empirical evidence that the REER in some commodity-producer countries is co-integrated with measures of relative productivity and/or terms of trade.

[8]Defined as the sum of sectoral productivity differentials (tradables and nontradables) across countries.

[9]In these models, improvements in terms of trade mainly affect the REER through an increase in nontradables prices, which result from higher wages in this sector following higher wages in the commodity-exporting sector in the aftermath of higher export prices in the presence of labor mobility. In Iran, most oil export earnings accrue to the budget. Thus, appreciation pressures following higher oil prices will stem mainly from expectations and improved business confidence if the underlying fiscal stance does not change.

In this empirical study on Iran, productivity differentials are omitted from the co-integration analysis for two reasons. First, available measures of productivity differentials with trading partners are not fully compatible with the theoretical formulations presented above and thus cannot be included in tests. This said, these indirect measures point to a lack of convincing evidence of higher productivity growth in Iran relative to trading partners. In fact, labor productivity in Iran grew at about 2 percent per year on average during 1993–2003, which is close to the weighted average productivity growth in trading partner countries (Appendix 2). Low, or even negative, TFP growth during 1993–2003 (Chapter 1) corroborates the results presented in Appendix 2. Second, there is no reliable high-frequency measure of relative productivity for Iran. Indeed, using annual data instead of monthly series will preclude any meaningful co-integration analysis.[10]

A real price of oil was used as a proxy for terms of trade, in line with similar research on other commodity producers (Cashin, Céspedes, and Sahay, 2002). The index of the real oil price was calculated using the U.K. Brent spot price deflated by the manufactured exports unit price index for developed countries.[11]

A visual inspection of REER1, REER2, and the real oil price series reveals two features. First, there is no clear pattern of co-movement of either REER series with the real oil price, in particular during 1997–98 (Figure 12). Second, volatility of both REER series appears to be lower than that of the real oil price.

These observations are confirmed by econometric tests. There is no co-integration between either REER1 or REER2 and the real oil price,[12] implying that the latter cannot explain the long-term REER dynamics. The standard deviation of REER1 and REER2 is 0.2, and that of the real oil price is 0.3.[13] This finding is surprising, because in most countries, including

[10]In many studies, relative productivities are measured indirectly as differentials between tradables and nontradables price indices, assuming a competitive equilibrium framework and market pricing mechanisms (Alberola and others, 1999). Such an approach is not relevant for Iran, because relative prices are not likely to reflect relative productivities, given the extensive price controls, widespread subsidies, and restrictive trade policy in effect for most of the period under review.

[11]On the basis of ADF test statistics with *p*-values in brackets— –1.17 (0.68)—the null hypothesis of nonstationarity of the real oil price index for 1993–2004 cannot be rejected. This finding is in line with other studies that have used longer time series and more sophisticated techniques (Cashin, Liang, and McDermott, 2000).

[12]The Aikaike Information Criterion and Schwartz Criterion indicate that one lag is appropriate. There is no co-integration for a system with one lag. Tests were also performed for systems with up to eight lags with no evidence of co-integration.

[13]Brown-Forsythe test (with *p*-values in brackets)—52.7 (0.0) and 35.7 (0.0)—confirms that the variances of REER1 and REER2, respectively, are not equal to those of the real oil price.

Figure 12. REER, M2, and Real Oil Price, 1993–2003
(Log 1999/2000 = 1)

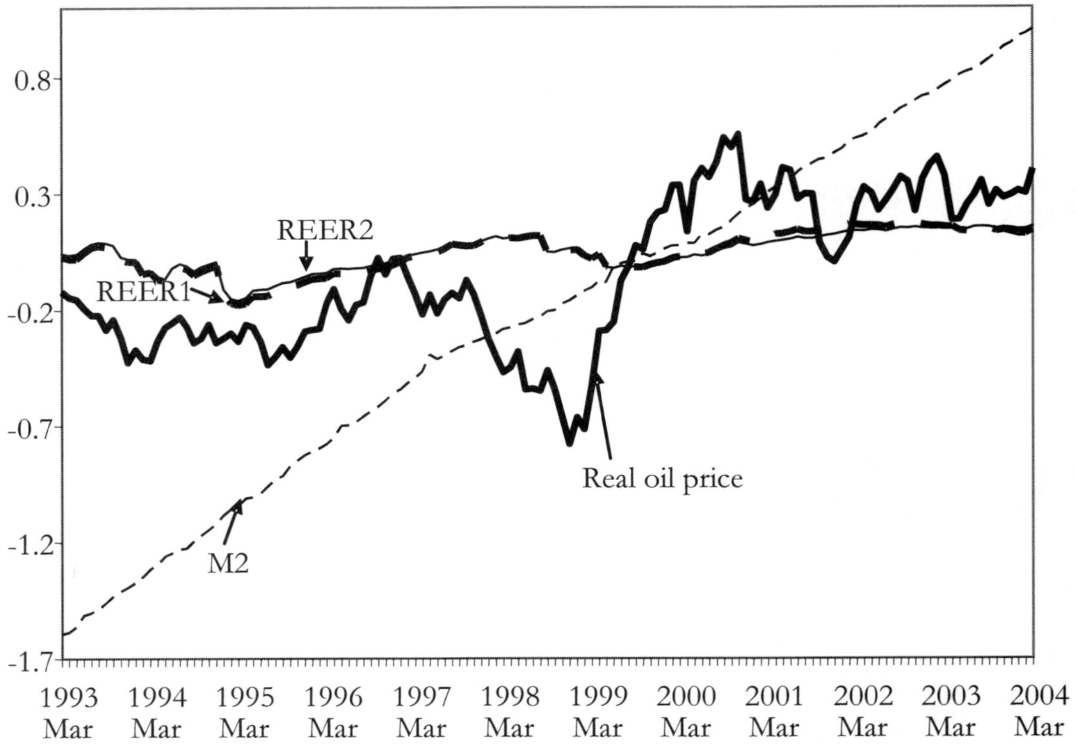

Sources: Iranian authorities; IMF staff estimates.

commodity producers, the REER is more volatile than fundamentals (MacDonald, 1999). This suggests that in Iran the market-related exchange rate was "managed" through intervention in the foreign exchange market and trade and payment restrictions, because the REER was less volatile than economic fundamentals.

The REER and Monetary Variables

The previous subsection shows that there is no statistically proven co-movement between the REER and the terms of trade. Thus, other factors that have a long-term impact on the REER should be considered. Monetary variables are potential candidates. In most theoretical settings, money has some short-term impact on real variables in open economy models, but it is neutral in the long run. But given mixed evidence behind the long-term neutrality of money, Obstfeld and Rogoff (1996) developed a model demonstrating that monetary shocks have short-term real effects due to sticky prices, and that they may also lead to temporary current account imbalances that have long-run wealth effects on the real exchange rate. However, they show that "the long-run effects must be smaller than the short-run effects" (page 681). Given these considerations, M2 was included in the co-integration relationship along with the terms of trade to explain the long-term dynamics of the REER in Iran. However, because in Iran fiscal policy directly affects monetary conditions (see Chapter 3), fluctuations of money supply will be a measure of monetary shocks and indirectly a proxy for the fiscal policy stance.[14]

Separate co-integration tests were conducted for REER1 and REER2. Other variables in both tests included the real oil price (LRPOIL) and seasonally adjusted M2 (LM2).[15] A model with one lag and a constant in the co-integration vector was selected in both cases.[16] The Johansen trace and eigenvalue tests indicate the existence of at least one co-integrating vector in both cases. The estimated co-integrating vector β is presented in equations (4) and (5):

$$\text{LREER1} = -0.14\ \text{LRPOIL} + 0.30\ \text{LM2} \quad (4)$$

$$\text{LREER2} = -0.15\ \text{LRPOIL} + 0.30\ \text{LM2} \quad (5)$$

The next step is to test whether long-term elasticity coefficients in β are different from zero. A restriction on long-term elasticity of money being equal to zero is rejected.[17] Thus, according to equations (4) and (5), higher money supply causes the REER to appreciate in the long run. As noted in Chapter 3, Iranian monetary expansions (contractions) broadly coincided with fiscal expansions (contractions), thus making REER dynamics indirectly related to the fiscal policy stance through money supply. This finding suggests that the fiscal impulse accommodated by monetary policy easing initiated in 2000 has contributed to the recent REER appreciation.

---

[14]Using fiscal variables directly is difficult, owing to a lack of reliable high-frequency fiscal data.

[15]All variables are in logarithms. Seasonal adjustment was made using the U.S. Census X12 package (multiplicative adjustment).

[16]Based on Aikaike and Schwartz information criteria.

[17]The restrictions are rejected at zero percent ($\chi^2 = 20.0$) and zero percent ($\chi^2 = 22.0$) confidence levels for equations (4) and (5), respectively.

The long-term impact of the real oil price on either REER1 or REER2 was not supported by the findings.[18] Thus, the issue of an equilibrium REER value consistent with the *real-side* long-term fundamentals is unresolved. There is no empirical support for the hypothesis that the substantial real appreciation registered by March 2004 relative to the 10-year mean also stems from the positive terms of trade shock that occurred at that time, partly due to the short data span and the various forms of government intervention.

The REER and the Tradables Sector

Given the limitations of the econometric techniques, a qualitative assessment of competitiveness is all the more important. Growth of non-oil exports and output performance of import-competing sectors provide good gauges of the current state of competitiveness. On these measures of competitiveness, Iran has performed well during 1999/2000–2003/04. Real value added in the manufacturing sector and non-oil exports grew at double-digit rates in many years (Figure 13).

It is difficult to reconcile such a good performance in the tradables sector with the appreciation in the REER. Since there is no clear evidence of strong improvement in productivity differentials, the explanation is likely to be related to the operating environment of the tradables sector. The magnitude of pricing advantages stemming from trade barriers and the cost savings resulting from subsidies are likely to be much larger than the fluctuations in the REER. For example, imports of cars were previously prohibited and then were subject to a 150 percent import tariff. Tariffs on other imports that compete with domestic products are still high, despite recent efforts toward trade liberalization. More important, exporters, as well as import-competing industries and agriculture, continue to benefit from tax exemptions, explicit subsidies, and negative real rates of charge on banking loans as well as heavily subsidized domestic energy prices.[19] The latter factor is particularly important for petrochemical exports. Finally, easier access to imports of capital and intermediate goods, as well as a reduction in tariffs on these categories, has recently improved competitiveness in the tradables sector.

---

[18]The restrictions cannot be rejected at 34 percent ($\chi^2 = 0.9$) and 25 percent ($\chi^2 = 1.3$) confidence levels for equations (4) and (5), respectively.

[19]For further details, see Chapters 1, 2, and 5.

Figure 13. Real Annual Growth Rates, 1993/94–2003/04
(In percent)

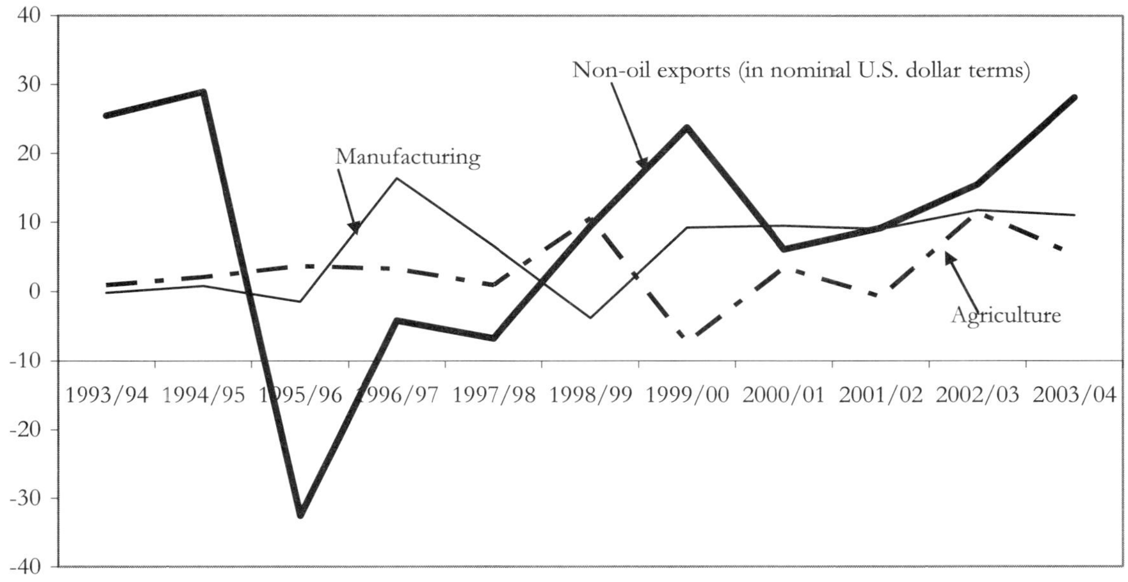

Sources: Iranian authorities; IMF staff estimates.

Forward-Looking Measures of the REER and the Current Account Norm

Since the exchange rate plays a key role in shaping external balances, the exchange rate level should be assessed in the context of an analysis of the savings-investment balance as well. The latter is usually discussed in economic literature from a forward-looking perspective (Obstfeld and Rogoff, 1996), and thus, the appropriate exchange rate level should also be consistent with the notion of an equilibrium intertemporal allocation of resources. In simple terms, if the observed current account balance is believed to be close to its long-term equilibrium level, the exchange rate can be considered to be close to equilibrium.

This chapter uses a normative approach to determine the appropriate ("equilibrium") level of the current account balance. Chinn and Prasad (2003) conducted an empirical study that "provides an indication of the levels of current accounts that may be considered 'normal' for a country, based on a number of its macroeconomic attributes" (page 48). In panel regressions for

Table 15. Normative Current Account Balance, 2004

	Panel Regression[1]		Panel Regression[2]	
	Five-year average	Regression coefficient	2003/04	Regression coefficient
Government balance[3]	0.015	0.259	–0.002	0.167
NFA/GDP[3]	0.078	0.039	0.098	0.036
Dependency ratio[4]	0.704	–0.055	. . .	. . .
M3/GDP[3]	0.496	0.037	. . .	. . .
Standard deviation of terms of trade	0.166	0.034	. . .	. . .
Relative income[5]	. . .	. . .	0.163	–0.049
Relative income squared[5]	. . .	. . .	0.027	0.114
Current account balance (–1)[3]	. . .	. . .	3.06	0.536
Normative current account balance[6]	0.0	. . .	1.6	. . .
Memorandum item:				
Current account balance in 2003/04[6]	1.5	. . .	1.5	. . .

Sources: Chinn and Prasad (2003); IMF staff estimates.
[1]Panel regression for developing countries, excluding Africa, OLS specification with time effects.
[2]Panel regression for emerging markets, excluding Africa, for annual data, OLS specification with time effects.
[3]As a fraction of GDP.
[4]Ratio of 0–14 year-olds to 15–64 year-olds.
[5]Ratio to the U.S. per capita income in purchasing power parity (PPP)-adjusted terms.
[6]Ten percent of GDP.

developing countries, excluding Africa, Chinn and Prasad (2003) determined the following macroeconomic attributes that affect the level of the current account balance: the ratio of the government fiscal balance to GDP, the ratio of net foreign assets to GDP, demographic dependency ratios,[20] terms of trade volatility, financial deepening, and relative PPP-adjusted incomes to the United States.[21] A norm for the current account balance in Iran was derived using the results of a number of panel regressions presented by Chinn and Prasad. Depending on the specification, the current account balance norm estimates for Iran range from a small deficit to a surplus of 1½ percent of GDP for 2003/04 (Table 15). Because the estimated norm for the current account balance is reasonably close to the actual realization, the REER level

[20]The ratio of the population younger than 15 years old to the working-age population (15–64 years old).
[21]A dummy for oil-producing countries was insignificant in the panel regressions.

appears broadly consistent with the "normal" level of the saving-investment balance, conditional on the fiscal stance in 2003/04 and other parameters.

Policy Implications of Findings

The previous sections have shown the difficulty of determining an equilibrium REER value based on real *and* monetary variables. However, other elements of the analysis do not point to a large misalignment of the exchange rate. Because monetary and fiscal policies are important determinants of long-term REER dynamics, fiscal and monetary tightening, in the context of limited capital mobility, could help reduce the speed of appreciation of the underlying unobserved equilibrium value of the REER in the short term, which may also stem from the recent increases in oil prices. This policy mix will contribute to faster growth of the non-oil sector by containing REER appreciation.

Conversely, trying to engineer a nominal depreciation of the effective exchange rate while pursuing an expansionary fiscal policy might lead to a further appreciation of the unobserved equilibrium value of the REER through an increase in inflation and even a short-term real exchange rate appreciation that exceeds this value.

Looking beyond the short term, such factors as oil prices, reform-driven productivity gains, progress in trade liberalization, subsidy reforms, and an eventual full capital account liberalization could have a significant impact on the equilibrium REER. However, given the uncertainties surrounding future developments in these fundamental factors and their impact on the equilibrium REER, the task of exchange rate management would become even more challenging. Thus, increased flexibility in exchange rate management might be needed, which would give the market an increasingly important role in determining the exchange rate. The central bank would have to tolerate a greater degree of two-way nominal exchange rate movements, while focusing on achieving its monetary policy objectives. The benefits of greater exchange rate flexibility can only be reaped if the monetary policy framework is substantially strengthened (Chapter 3).

Conclusion

A judgment on the appropriate level of the REER in Iran can be based on a combination of considerations. The PPP does not appear to be a useful yardstick for exchange rate management. More refined econometric analysis does not appear to yield significant results. Although monetary policy and the underlying fiscal policy stance appear to have a long-term impact on the REER, the terms of trade do not appear to explain the long-term evolution of the market-related REER. Thus, the extent to which recent REER appreciation is justified by terms of trade developments cannot be inferred from the data sample.

Given the limitations of the econometric analysis, the REER level was analyzed in light of the performance of the tradables sector and the current account norm estimated for Iran based on cross-country panel data. Owing to good performance of the tradables sector and the proximity of the current account balance to the "norm," the exchange rate level does not appear to have been misaligned. This conclusion is subject to important caveats regarding various distortions that have helped improve the performance of Iran's tradables sector, but at the expense of efficient allocation of resources, which might undermine future growth performance.

The short- and medium-term challenge facing the authorities is to avoid a sharp real appreciation of the unobserved equilibrium level of the REER, which could be triggered by a combination of high oil prices and expansionary policies. In the longer term, greater flexibility in exchange rate management is needed in the face of uncertainties related to the external environment and the effects of important structural transformations envisaged in the authorities' reform agenda.

CHAPTER

5 Issues in Medium-Term Management of Oil Resources

Under the Third Five-Year Development Plan (2000/01–2004/05), Iran introduced a number of important fiscal reforms aimed at reducing the dependency of public finances on oil revenue and containing expenditure growth. These included a tax reform to strengthen the non-hydrocarbon revenue base, the establishment of the Oil Stabilization Fund (Appendix 3) to cushion the impact of fluctuations in oil prices on expenditure, and actions to improve expenditure management and transparency.

Despite these efforts, fiscal reform in general has been limited, and measures have often been implemented under pressing domestic and external political circumstances. Moreover, increases in oil prices have provided temporary respites from underlying long-term fiscal problems and have given policymakers more leeway to increase expenditure. As a result, Iran's public finances continue to face a number of key challenges: heavy dependence on hydrocarbon revenue, low non-hydrocarbon revenue, procyclical fiscal policy, and excessive subsidization (Appendix 4). These issues need to be addressed in a medium-term framework, which would also help smooth the macroeconomic impact of fluctuations in oil prices, reduce inflation, and preserve hydrocarbon wealth for future generations.

This chapter outlines long-term objectives for preserving hydrocarbon wealth for future generations and discusses a medium-term fiscal outlook that is consistent with these objectives and with the need to bring down inflation and reduce vulnerability to a possible decline in oil prices. The discussion is focused on central government operations consolidated with the OSF, mainly because data limitations preclude an analysis of general government operations, including municipal budgets and the Social Security Organization.[1] This chapter does not address the issues of quasi-fiscal

---

Note: Prepared by José Bailén and Vitali Kramarenko.

[1]According to preliminary studies, the Social Security Organization is not expected to face any cash flow problems owing to positive demographic dynamics for the next 10 to 15 years. Beyond this period, deficits are expected to emerge, but the exact amount of unfunded liabilities is not available.

activities in the banking system (Chapter 2), fiscal dominance over monetary policy (Chapter 3), or the relationship between public enterprises and central government financial operations.

The next section of the chapter derives three sets of long-term simulations of consumption from hydrocarbon wealth. Then the results of these simulations are compared with a baseline medium-term fiscal projection in order to assess whether a transition from the projected baseline scenario toward a fiscal position suggested by the simulations is feasible.[2] The chapter concludes with recommendations for incorporating long-term considerations into the design of the medium-term fiscal framework, with an emphasis on maintaining real per capita hydrocarbon wealth constant in the medium term.

Long-Term Considerations

Analytical Framework for Long-Term Analysis

In oil-producing economies, fiscal policy should aim to accumulate substantial net assets during the period of oil production to cover non-oil deficits in the post-oil period (Barnett and Ossowski, 2003). This section draws on an intertemporal optimization framework[3] to estimate optimal government consumption and savings from hydrocarbon wealth.[4] In this context, intergenerational equity considerations are given prominence, while fiscal sustainability issues are not explicitly examined, assuming that the government's intertemporal budget constraint is always met.[5] The theory stipulates that the optimal consumption path is a function of the net present value of oil revenue, the initial net debt of the government, the real rate of return on assets, an intertemporal discount factor, the rate of population growth, and the degree of risk aversion (Box 4).

---

[2]The baseline scenario is provided for illustrative purposes only and does not represent the staff's current projections. The medium-term scenarios discussed in this chapter do not take into account the changes in oil market conditions and prospects that have occurred since 2004.

[3]Engel and Valdés (2000) provide an overview of the application of optimal consumption models to the analysis of fiscal sustainability in oil-producing countries.

[4]See also Azerbaijan (Wakeman-Linn and others, 2004) and Kazakhstan (Davoodi, 2002).

[5]The intertemporal budget constraint requires that hydrocarbon wealth be equal to the present value of future current primary balances. All simulations in this paper assume that any estimated decline in consumption from oil wealth relative to GDP will be matched by either revenue or expenditure measures to maintain the intertemporal budget constraint. The issue of medium-term fiscal sustainability is not addressed in this paper, because it is a complex concept involving the analysis of the composition of government net debt by maturity, currency and instruments, and financing constraints.

Box 4. Optimal Consumption from Hydrocarbon Wealth

The purpose of this model is to determine an optimal rule on how to distribute hydrocarbon wealth across generations. The optimal solution to the government's consumption level is derived from equation (1):

$$U = 1/(1-p) \sum \beta^{t} C_{G,t}^{1-\rho}, \quad (1)$$

where:

U is welfare function;

β is a discount factor of the social planner in the welfare function; $\beta = 1/(1-\gamma)$ where γ is a discount rate; and

$C_{G,t}$ is government consumption at time t.

Equation (2) defines net wealth $W_{G,0}$ as the starting net wealth $F_{G,0}$ and the present discounted value of future oil revenues $Y_{G,s}$:

$$W_{G,0} = F_{G,0} + \sum (1+r)^{-s} Y_{G,s}, \quad (2)$$

where:

r is a real rate of return on assets, and

$Y_{G,s}$ is oil revenue in period s.

Equations (3) and (4) define the optimal path of government current spending from hydrocarbon wealth:

$$C_{G,0} = (1-\alpha)(1+r) W_{G,0}, \quad (3)$$

$$C_{G,t+1} = [\beta(1+r)]^{1/\rho} C_{G,t}, \quad (4)$$

$$\alpha = (1+n)\,[\beta(1+r)]^{1/\rho} / r,$$

where:

$1/\rho$ denotes the coefficient of intertemporal substitution of consumption between two periods or the coefficient of relative risk aversion in a stochastic framework; and

n is the population growth rate.

If:

(a) $\beta(1+r) > 1$. If the welfare of future generations is discounted at 1 percent (implying $\beta = 0.99$), consumption grows at $C_{G,t+1} / C_{G,t} = [\beta(1+r)]^{1/p}$. It is straightforward that about 3 percent growth of consumption is consistent with $\rho = 1.5$ if $\beta = 0.99$ and $R = 0.04$.

(b) $\beta(1+r) = 1$. Because the discount rate is equal to the real rate of return, the right-hand side of equation (3) is simplified as:

$$C_{G,0} = (r-n)\, W_{G,0}. \quad (5)$$

(c) If $\beta(1 + r) < 1$, there will be a decline in consumption growth no matter what ρ might be.

The paper presents three sets of simulations for consumption from hydrocarbon wealth. The first simulation assumes that the discount rate in the welfare function is below the real rate of return. This is consistent with positive long-term growth of per capita output and consumption and implies that the society is patient and saves enough to ensure real per capita growth in consumption from hydrocarbon wealth. The second simulation assumes that the discount rate is equal to the real rate of return. Under this assumption, there is no long-term real per capita growth and, therefore, real per capita consumption from hydrocarbon wealth will remain constant indefinitely, which is akin to the conclusions of the permanent income hypothesis. In the third simulation, a preservation of real hydrocarbon wealth is targeted. This is equivalent to assuming that the discount rate is higher than the real rate of return, given the positive rate of population growth.

The issue of whether savings are invested in financial or real (physical) assets does not affect the general conclusions of the optimal consumption theory. Public investment in infrastructure and human capital financed by savings from oil revenue can contribute to an increase in the long-term growth rate of the non-oil sector. Fiscal sustainability, however, requires that public investment be sufficiently productive to generate tax revenue higher than or equal to the prevailing return on financial assets of the equivalent amount.[6] Assuming that this rule is observed, the following analysis does not make a distinction between investment in financial and physical assets.

As a share of oil revenue is saved and invested, the return on these investments becomes an important source of additional non-oil revenue for the budget. Accordingly, the government's consumption from hydrocarbon wealth can be measured by the non-oil current deficit, including implicit energy subsidies as current expenditure (Table 16), minus net interest income.[7]

Estimates of the path of optimal consumption from hydrocarbon wealth are subject to significant uncertainties. They are highly sensitive to several factors, including long-term oil and gas prices, the volume of proven reserves, the extraction rate, and the real rate of return. Of these factors, oil and gas prices are the most difficult to predict. Some empirical research suggests that oil prices do not revert to a long-term mean, whereas other research finds only a very slow reversion and high persistence of shocks (Cashin, Liang, and McDermott, 2000; and Engel and Valdés, 2000). Uncertainties regarding the oil revenue outlook may also stem from the

[6]This principle is valid regardless of the presence of oil resources.

[7]In this chapter, this definition is used for non-oil primary current deficit.

Table 16. Estimates of Implicit Energy Subsidies

	2003/04
Domestic sales prices in rials	
Kerosene (rial/liter)	160
Fuel oil (rial/liter)	88
Gasoline (rial/liter)	713
Gas oil (rial/liter)	160
Electricity (rial/kwh)	130
Natural gas (rial/m^3)	71
Border prices in rials (at market exchange rates)	
Kerosene (rial/liter)	1,187
Fuel oil (rial/liter)	676
Gasoline (rial/liter)	1,696
Gas oil (rial/liter)	1,131
Electricity (rial/kwh)	421
Natural gas (rial/m^3)	410
Implicit subsidy (in trillions of rials)	117
Kerosene	11
Fuel oil	5
Gasoline	18
Gas oil	25
Electricity	33
Natural gas	26
Memorandum items:	
Implicit subsidy in percent of GDP	10.4
Implicit subsidy in billions of dollars	14.2
Average exchange rate (rials per U.S. dollar)	8,282.0

Sources: World Bank; IMF staff estimates.

Table 17. Parameters of Oil and Gas Wealth Estimates

Proven oil reserves (in billions of barrels)	130.7
Proven gas reserves (in trillions of m^3)	26.7
Oil production in 2002/03 (in millions of barrels per day)	3.2
Gas production in 2002/03 (in billions of m^3)	100.0
Long-term real price for Iranian crude oil (at constant 2003/04 dollars per barrel)	22.0
Long-term real price of Iranian gas (at constant 2003/04 dollars per m^3)	41.0
Long-term annual (real) return on capital (percentage)	4.0
Expected average annual rate of population growth (percentage)	1.1
Government external net debt (in billions of dollars) at end-2003/04	7.3

Sources: World Bank; British Petroleum Statistical Bulletin; IMF staff estimates and projections.

discovery of additional oil and gas reserves or the development of alternative energy sources. Thus, the estimated consumption path needs to be frequently revisited, because new information may lead to large variations in estimates.

Implications for Iran of Long-Term Analysis

The estimate of hydrocarbon wealth hinges on a number of key assumptions. The assumed oil price of $22 per barrel is equal to a 10-year average oil price for Iranian crude (1993–2003) in 2003/04 U.S. dollars. Because of the market segmentation for natural gas, measuring border prices for Iranian gas is a major challenge. A conservative assumption of $41 per 1,000 cubic meters was used in this scenario; this figure is somewhat lower than the 10-year average prices in Europe or North America. On the basis of proven reserves, the country's oil and gas resources are estimated to last for 75 and 78 years, respectively, assuming that extraction of oil grows at 1 percent per year and extraction of gas accelerates in the next decade and then levels off. A real rate of return of 4.0 percent is assumed, broadly in line with long-term real U.S. Treasury bond rates of 2 percent (60 percent weight) and real stock returns of 7 percent (40 percent weight).[8] Government domestic and external debt, net of OSF foreign exchange deposits, is estimated at about $10 billion. Given the assumptions in Table 17, the overall oil and gas wealth, net of government debt, is about $861 billion in 2003/04 dollars (Table 18).

[8]These weights are chosen for illustrative purposes only. The optimal weights should be determined based on a capital asset pricing model.

Table 18. Consumption from Oil Wealth

(In billions of current U.S. dollars, unless otherwise indicated)

	Preliminary 2003/04	Projected 2004/05
Total	860.8	. . .
Oil wealth	645.1	. . .
Gas wealth	225.8	. . .
Initial debt, net of OSF foreign exchange deposits	10.0	. . .
Estimated consumption from oil wealth if:		
Oil wealth constant in real terms	34.4	35.1
Oil wealth constant in real per capita terms	25.2	25.9
Optimal consumption from oil wealth	8.0	8.4
Estimated consumption from oil wealth (in percent of GDP) if:		
Oil wealth constant in real terms	25.2	22.2
Oil wealth constant in real per capita terms	18.4	16.4
Optimal consumption from oil wealth	5.9	5.3
Estimated consumption:	29.5	33.8
In percent of GDP	21.6	21.4
Non-oil primary balance, excluding capital expenditure	15.3	17.2
Implicit subsidies	14.2	16.6

Sources: IMF staff estimates and projections.

Simulation of optimal consumption from oil wealth. According to equation (4) in Box 4, the optimal consumption from hydrocarbon wealth is estimated at $8 billion in 2003/04, compared with the estimated consumption of about $29 billion. Under this framework, the optimal per capita consumption from hydrocarbon wealth should increase every year, which would help maintain the amount of consumption from hydrocarbon wealth roughly constant in terms of GDP at about 5 percent indefinitely (assuming 3 percent real GDP growth in the long run, as in Figure 14).

Figure 14. Optimal Path for Consumption from Oil Wealth, 2003–83

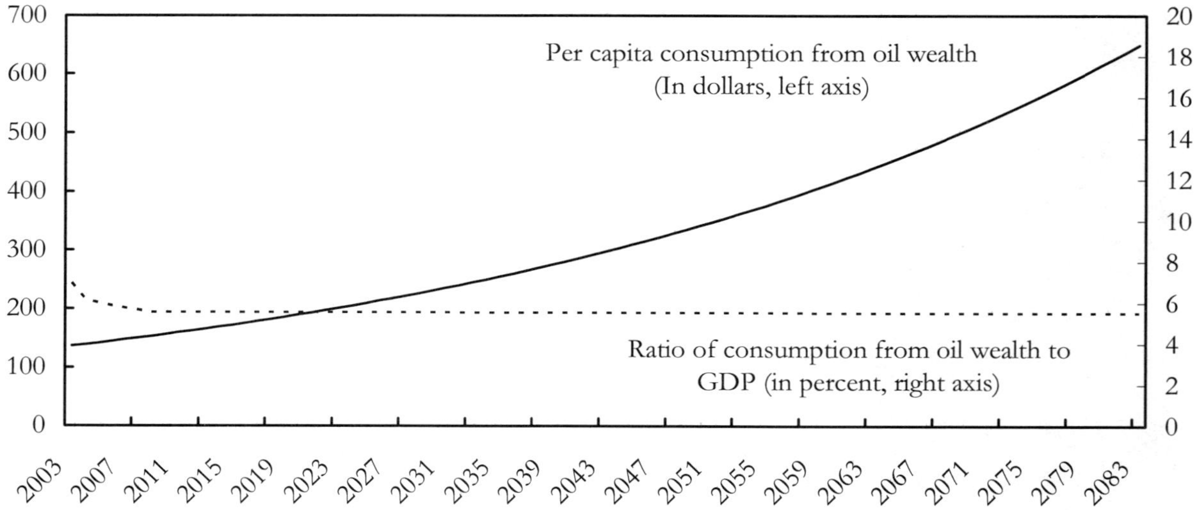

Source: IMF staff estimates and projections.

Figure 15. Constant per Capita Real Consumption from Oil Wealth, 2003–83

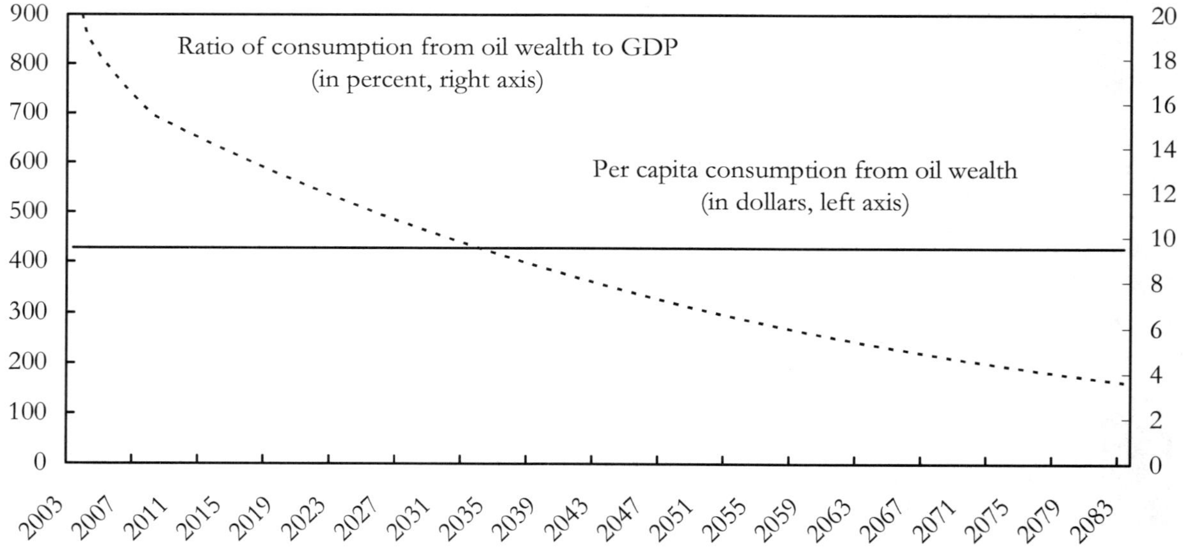

Source: IMF staff estimates and projections.

Simulation of consumption from oil wealth consistent with maintaining constant real per capita wealth. To maintain hydrocarbon wealth constant in real per capita terms, the limit on consumption from hydrocarbon wealth is estimated at $25 billion in 2003/04. Under this framework, such consumption will increase in constant dollar terms every year at the rate of population growth but will decline rapidly as a share of GDP, because the rate of population growth of 1 percent is well below the assumed long-term GDP growth rate of 3 percent (Figure 15). Thus, to maintain the intertemporal budget constraint in the long run, the current primary deficits would need to decline in the future, which would require some additional fiscal measures.

Simulation of consumption from oil wealth consistent with maintaining constant real wealth. To maintain hydrocarbon wealth constant in real terms, the real return ($34 billion) on the total wealth could be consumed from 2003/04 onward. The long-term implication of this assumption is that consumption from hydrocarbon wealth both in real per capita terms and relative to GDP would decline steeply, potentially warranting sharp increases in taxation or reductions in expenditure to maintain the intertemporal budget constraint. Another concern associated with this scenario (which is not captured by the framework) is a possible substantial REER appreciation that could result from the large spending out of foreign-currency-denominated oil revenue. If sustained, such appreciation could cause Dutch disease which would hinder the development of the non-oil sector.[9]

Sensitivity to oil price assumptions. Only under the simulation of the optimal policy scenario can consumption from hydrocarbon wealth be indefinitely maintained constant relative to GDP. The corresponding current primary non-oil deficit, estimated at about 5 percent of GDP, is sensitive to assumptions about oil prices. Should the assumed real oil prices be higher (lower) during the projection period, the long-term optimal consumption from hydrocarbon wealth would increase (decrease) by about ¼ percent of GDP for each dollar in excess of (below) the baseline oil price of $22 per barrel. For instance, if the assumed real oil price is $30 per barrel, the optimal level of consumption from hydrocarbon wealth is 6.8 percent of GDP, or about 2 percent of GDP above the level of consumption consistent with the baseline price of $22 per barrel. The government could indefinitely maintain consumption from hydrocarbon wealth at its 2003/04 level relative to GDP (22 percent) only if the assumed real oil price were about $88 per barrel. In

[9]Dutch disease describes a situation in which, as the real exchange rate appreciates, the tradable goods sector contracts compared with the nontradables sector because the currency appreciation makes tradable goods less competitive and leads to an increase in imports. The result is a shift of resources away from the production of tradable goods and toward nontradables.

the scenario to preserve real per capita wealth, consumption from oil wealth is estimated to increase (decrease) by $1 billion in 2003/04 for each dollar in excess of (below) the baseline price. With respect to maintaining total real oil wealth, consumption from oil wealth is estimated to increase (decrease) by almost $2 billion in 2003/04 for each dollar in excess of (below) the baseline price.

Consistency of the Baseline Medium-Term Framework with Long-Term Parameters

A stylized baseline medium-term fiscal scenario is developed as a reference for comparison with the three simulations presented above (Figure 16). This scenario is based on a number of assumptions on reforms contained in the draft Fourth Five-Year Development Plan, including fiscal reforms. This scenario is also based on a conservative assumption of a steady decline in oil prices to about $24 per barrel at the end of the plan period, as well as a gradual fiscal adjustment that will be achieved mainly through additional revenue measures, including energy subsidy reform[10] and value-added tax (VAT) implementation in 2006/07. A gradual reduction in explicit subsidies is also assumed.

All three sets of long-term simulations point to the need for a change in the fiscal stance under the baseline scenario. This scenario is somewhat tighter than is needed to preserve hydrocarbon wealth in real terms over the next five years. Maintaining real per capita hydrocarbon wealth constant calls for additional moderate fiscal tightening relative to the baseline scenario (see Figure 15). Shifting the fiscal stance toward the optimal consumption path will necessitate a drastic fiscal tightening relative to the baseline scenario, which is clearly not feasible over the medium term.

Although a transition toward an optimal consumption path would result in maintaining consumption from hydrocarbon wealth constant relative to GDP indefinitely,[11] such a transition can only be accomplished in stages over the long run. As a first step, the preservation of real per capita hydrocarbon wealth could be achieved over the medium term; subsequently, consideration could be given to moving closer to maintaining constant consumption from oil wealth relative to GDP.

---

[10]On the basis of broad indications regarding the planned energy price reforms, it is assumed that energy prices will increase by 25 percent per year and that two-thirds of the additional revenue generated by the reform will be spent on social protection and public investment.

[11]Barnett and Ossowski (2003) recommend this criterion as a benchmark of long-term fiscal sustainability in oil-producing countries.

Figure 16. Scenarios for Consumption from Oil Wealth, 2003/04–2009/10
(In percent of GDP)

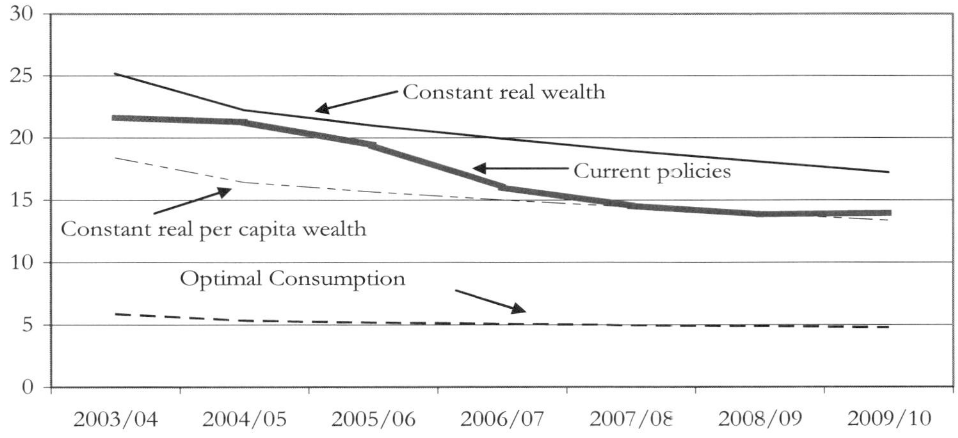

Source: IMF staff estimates and projections.

To maintain oil wealth constant in real per capita terms, consumption from oil wealth under the baseline scenario would have to be reduced by about 4 percentage points of GDP during 2004/05–2005/06, with a smaller adjustment of about 0.5 percentage point of GDP during 2007/08–2009/10. Given the need to reduce external and fiscal vulnerability to a potential decline in oil prices, an up-front fiscal adjustment is preferable, because it is less costly to implement from a position of strength when oil prices are high. Indeed, the current fiscal impulse[12] has contributed to a significant decline in the current account surplus despite higher oil prices, making the external position more vulnerable to a possible decline in oil prices in the medium term. However, a fiscal tightening of about 4 percentage points of GDP per year at the beginning of the projection period, relative to the baseline scenario, appears excessive, because it could have an unacceptably high output cost.

Conclusion

This chapter relies on a long-term analytical framework to determine criteria for preserving hydrocarbon wealth for future generations. It presents three possible long-term criteria in comparison with a medium-term baseline fiscal

[12]The non-oil fiscal deficit increased from 10.8 percent of GDP in 1999/2000 to 18.7 percent in 2002/03; it declined to 16.4 percent of GDP in 2003/04.

scenario: (1) maintaining total real hydrocarbon wealth constant, (2) preserving real per capita hydrocarbon wealth, and (3) targeting optimal consumption from hydrocarbon wealth. The baseline fiscal scenario is broadly in line with maintaining total real hydrocarbon wealth constant. However, additional fiscal adjustment relative to the baseline is needed to move toward preserving real per capita hydrocarbon wealth or converging with the optimal path for consumption from hydrocarbon wealth. Short- and medium-term factors are likely to play a crucial role in choosing the appropriate medium-term path of fiscal adjustment. Given the current fiscal stance, it is likely that a feasible option would lean toward the preservation of hydrocarbon wealth in real per capita terms rather than to the optimal path. This means that additional fiscal measures might need to be considered beyond the medium term to deal with the projected decline in consumption from hydrocarbon wealth relative to GDP.

APPENDIX

1 Data Sources and Methodology

The source for real GDP and investment data for Iran is the latest Central Bank of Iran database; for the rest of the countries, the source is the International Financial Statistics database. The source for employment data for 1960–90 is the International Labor Organization database (1956, 1966, 1976, and 1986 census), and the source for employment data after 1990 is the census conducted by the Statistical Center of Iran. The growth accounting exercise follows the methodology described in Barro and Sala-i-Martin (1995, Chapter 10). The capital stock depreciation rate is 4.9 percent, consistent with the estimates of the Central Bank of Iran, and the initial capital stock is determined through the "rough guess" method suggested by Barro and Sala-i-Martin (1995). The average annual rate of return of capital is 7 percent, which is the long-term international average rate of return estimated by Siegel (1998). The average years of schooling of the labor force are drawn from the Barro and Lee (2000) database for every five years and extrapolated for the periods within each five-year period. Human capital is estimated in terms of average years of schooling following the standard definition used by Lucas (1988).

APPENDIX

2 Productivity Growth in Iran and Its Trading Partner Countries, 1993–2002

Productivity Growth in Iran and its Trading Partner Countries, 1993–2002
(Average annual percentage change, unless otherwise indicated)

Country	Weights (in percent)	Annual Productivity Growth
Australia	1.89	2.09
Austria	2.76	1.99
Belgium	3.64	1.56
Brazil	1.62	–0.80
Canada	1.86	1.83
China	2.53	8.98
France	5.90	1.20
Germany	23.96	1.23
Hungary	1.21	4.59
Italy	9.48	1.54
Japan	14.74	1.39
Netherlands	4.48	0.75
Spain	2.31	0.01
Sweden	2.05	3.09
Switzerland	3.12	1.46
Turkey	5.41	2.67
United Kingdom	7.19	2.30
United States	5.87	1.96
Total for trading partners	100.00	1.73
Iran	. . .	1.70

Sources: IMF, WEO; IMF staff estimates.

APPENDIX

3 The Oil Stabilization Fund

The Oil Stabilization Fund (OSF) was established in December 2000 with the objective of insulating the budget from fluctuations in oil prices. The fund was established as a foreign currency account at the central bank and is managed by an executive committee composed of the minister of economy and finance, the head of the Management and Planning Organization, the governor of the central bank, and two members of parliament.

The Third Five-Year Development Plan established a U.S. dollar ceiling on the oil export revenue that can be transferred to the budget, based on an oil price of about $16 per barrel. Additional transfers must be approved by parliament. Oil revenues in excess of the budgeted amount are transferred to the OSF. If the realized crude oil export revenue is less than the annual budget allocation by the end of the eleventh month of the fiscal year, the central bank draws from the OSF the amount required to make up the shortfall and transfers its equivalent in rials to the treasury.

All OSF foreign assets are held in a foreign-currency deposit account at the central bank; no more than 50 percent may be lent out domestically in foreign currency through the domestic banking system to the private sector at rates of return close to the London interbank offered rate (LIBOR). A firm may borrow from the OSF over an eight-year period and is required to reimburse its loan from the fifth to the eighth year of the project. The required collateral for the loan may be land, machinery, equipment, and/or corporate bonds.

APPENDIX

4 Subsidies

Explicit subsidies include budgetary subsidies to households for essential commodities, such as wheat, rice, oil, sugar, milk, and cheese; imports of medical equipment and pharmaceuticals; fertilizers; and some debt service payments on publicly guaranteed debt. Most of these subsidies were provided implicitly through the official exchange rate until the 2002 exchange rate unification, when the subsidies started being explicitly estimated in the budget. Subsidies are also provided to farmers by a specialized government-owned agency to guarantee minimum purchase prices of agricultural products. Food subsidies are rationed through coupons, which are given to all households regardless of income level. In recent years, the government has gradually reduced explicit subsidies for food, although they still constituted about 4 percent of GDP in 2003/04. Subsidies to cover exchange rate losses on debt service are declining rapidly as debt contracted before the 1993 and 2002 exchange rate unifications comes to maturity.

Implicit energy subsidies have given rise to misallocation of resources, waste, and overconsumption of energy products. As a result, Iran has become one of the most energy-intensive countries in the world. Total oil consumption amounted to 1.5 million barrels per day in 2002/03—a rate similar to that of Spain, which has a GDP six times higher than Iran. Also, air pollution is emerging as one of the main environmental and health problems, especially in Tehran.

Bibliography

Alberola, Enrique, Susana G. Cervero, J. Humberto Lopez, Angel Ubide, 1999, "Global Equilibrium Exchange Rates: Euro, Dollar, 'Ins,' 'Outs,' and Other Major Currencies in a Panel Cointegration Framework," IMF Working Paper 99/175 (Washington: International Monetary Fund).

Alesina, Alberto, Sule Özler, Nouriel Roubini, and Phillip Swagel, 1996, "Political Instability and Economic Growth," *Journal of Economic Growth,* Vol. 1 (June), pp. 189–211.

Arestis, Philip, and Panicos Demetriades, 1997, "Financial Development and Economic Growth: Assessing the Evidence," *Economic Journal*, Vol. 107 (May), pp. 783–99.

Barnett, Steven, and Rolando Ossowski, 2003, "Operational Aspects of Fiscal Policy in Oil-Producing Countries," in *Fiscal Policy Formulation and Implementation in Oil-Producing Countries*, ed. by J.M. Davis, R. Ossowski, and A. Fedelino (Washington: International Monetary Fund).

Barro, Robert J., 1991, "Economic Growth in a Cross Section of Countries," *Quarterly Journal of Economics,* Vol. 106 (May), pp. 407–43.

———, 1996, "Democracy and Growth," *Journal of Economic Growth,* Vol. 1 (March), pp. 1–27.

———, 1997, "Determinants of Economic Growth: A Cross-Country Empirical Study," Institute for International Development Discussion Paper No. 579 (Cambridge, Massachusetts: Harvard University).

———, and Jong-Wa Lee, 2000, "International Data on Educational Attainment Updates and Implications," NBER Working Paper No. 7911 (Cambridge, Massachusetts: National Bureau of Economic Research).

———, and Xavier Sala-i-Martin, 1995, *Economic Growth* (New York: McGraw-Hill).

Benhabib, J., and M.M. Spiegel, 1994, "The Role of Human Capital in Economic Development: Evidence from Aggregate Cross-Country Data," *Journal of Monetary Economics,* Vol. 34 (October), pp. 143–73.

Bils, Mark, and Peter J. Klenow, 2000, "Does Schooling Cause Growth?" *American Economic Review*, Vol. 90 (December), pp. 1160–83.

Bleaney, M.F., 1996, "Macroeconomic Stability, Investment and Growth in Developing Countries," *Journal of Development Economics*, Vol. 48 (March), pp. 461–77.

Cashin, Paul, Luis Céspedes, and Ratna Sahay, 2002, "Keynes, Cocoa, and Copper: In Search of Commodity Currencies," IMF Working Paper 02/223 (Washington: International Monetary Fund).

Cashin, Paul, Hong Liang, and C. John McDermott, 2000, "How Persistent Are Shocks to World Commodity Prices?" *IMF Staff Papers*, Vol. 47 (December), pp. 177–217.

Celasun, Oya, 2003, "Exchange Rate Regime Considerations in an Oil Economy: The Case of the Islamic Republic of Iran," IMF Working Paper 03/26 (Washington: International Monetary Fund).

———, and Mangal Goswami, 2002, "An Analysis of Money Demand and Inflation in the Islamic Republic of Iran," IMF Working Paper 02/205 (Washington: International Monetary Fund).

Chen, Yu-chin, and Kenneth Rogoff, 2002, "Commodity Currencies and Empirical Exchange Rate Puzzles," IMF Working Paper 02/27 (Washington: International Monetary Fund).

Chinn, Menzie, and Eswar Prasad, 2003, "Medium-Term Determinants of Current Accounts in Industrial and Developing Countries: An Empirical Exploration," *Journal of International Economics*, Vol. 59 (January), pp. 47–76.

Creane, Susan, Rishi Goyal, A. Mushfiq Mobarak, and Randa Sab, 2003, *Financial Development in the Middle East and North Africa* (Washington: International Monetary Fund).

Davis, J.M., R. Ossowski, and A. Fedelino, eds., 2003, *Fiscal Policy Formulation and Implementation in Oil-Producing Countries* (Washington: International Monetary Fund).

Davis, Jeffrey, Rolando Ossowski, James Daniel, and Steven Barnett, 2001, *Stabilization and Savings Funds for Nonrenewable Resources: Experience and Fiscal Policy Implications*, IMF Occasional Paper No. 205 (Washington: International Monetary Fund).

Davoodi, Hamid, 2002, "Assessing Fiscal Vulnerability, Fiscal Sustainability, and Fiscal Stance in a Natural Resource Rich-Country," in *Republic of Kazakhstan: Selected Issues and Statistical Appendix*, IMF Country Report No. 02/64 (Washington: International Monetary Fund).

De Gregorio, Jose, and Holger Wolf, 1994, "Terms of Trade, Productivity, and the Real Exchange Rate," NBER Working Paper No. 4807 (Cambridge, Massachusetts: National Bureau of Economic Research).

Demetriades, P., and K. Hussein, 1996, "Does Financial Development Cause Economic Growth? Time-Series Evidence from 16 Countries," *Journal of Development Economics*, Vol. 51 (December), pp. 387–411.

Demetriades, Panicos O., Michael P. Devereux, and Kul B. Luintel, 1998, "Productivity and Financial Sector Policies: Evidence from South East Asia," *Journal of Economic Behavior and Organization*, Vol. 35 (March), pp. 61–82.

Dornbusch, Rudiger, 1987, "Purchasing Power Parity," in *The New Palgrave: A Dictionary of Economics*, ed. by John Eatwell, Murray Milgate, and Peter Newman (New York: Stockton Press).

Easterly, William, Michael Kremer, Lant Pritchett, and Lawrence H. Summers, 1993, "Good Policy or Good Luck?: Country Growth Performance and Temporary Shocks," *Journal of Monetary Economics,* Vol. 32 (December), pp. 459–83.

Engel, Eduardo, and Rodrigo Valdés, 2000, "Optimal Fiscal Strategy for Oil Exporting Countries," IMF Working Paper 00/118 (Washington: International Monetary Fund).

Fischer, S., 1993, "The Role of Macroeconomic Factors in Growth," *Journal of Monetary Economics,* Vol. 32 (December), pp. 485–512.

Greenaway, David, Wyn Morgan, Peter W.Wright, 1998, "Trade Reform, Adjustment and Growth: What Does the Evidence Tell Us?" *Economic Journal*, Vol. 108 (September), pp. 1547–61.

Hassan, Sabir Mohammed, 2002, "Toward Fostering Financial Stability for the Islamic Financial Industry," paper presented at the inauguration of the Islamic Financial Services Board, Kuala Lumpur, Malaysia.

Hinkle, Lawrence E., and Peter J. Monteil, 1999, *Exchange Rate Misalignment: Concepts and Measurement for Developing Countries* (Washington: Oxford University Press for the World Bank).

Ishii, Shogo, 2003, *Exchange Arrangements and Foreign Exchange Markets: Developments and Issues,* World Economic and Financial Surveys (Washington: International Monetary Fund).

———, and Karl Habermeier, 2002, *Capital Account Liberalization and Financial Sector Stability*, IMF Occasional Paper No. 211 (Washington: International Monetary Fund).

Jalali-Naini, Ahmad R., 2003, "Economic Growth in Iran: 1950–2000" (unpublished; IRPD, Tehran, Iran). Available via the Internet: http://www.erf.org.eg/html/grp_list%202.asp#Iran

Jbili, Abdelali, and Vitali Kramarenko, 2003, *Choosing Exchange Regimes in the Middle East and North Africa* (Washington: International Monetary Fund).

Khan, Mohsin S., and Abdelhak S. Senhadji, 2001, "Threshold Effects in the Relationship Between Inflation and Growth," *IMF Staff Papers,* Vol. 48 (October), pp. 1–21.

Kwiatkowski, Denis, Peter Phillips, Peter Schmidt, and Yongcheol Shin, 1992, "Testing the Null Hypothesis of Stationarity Against the Alternative of a Unit Root," *Journal of Econometrics*, Vol. 54 (December), pp. 159–78.

Leeper, E.M., 1991, "Equilibria Under 'Active' and 'Passive' Monetary and Fiscal Policies," *Journal of Monetary Economics*, Vol. 27 (February), pp. 129–47.

Levine, Ross, 1997, "Financial Development and Economic Growth: Views and Agenda," *Journal of Monetary Economics*, Vol. 35 (June), pp. 688–726.

Lucas, Robert E. Jr., 1988, "On the Mechanics of Economic Development," *Journal of Monetary Economics*, Vol. 22 (July), pp. 3–42.

MacDonald, Ronald, 1999, "Exchange Rate Behaviour: Are Fundamentals Important?" *The Economic Journal*, Vol. 109 (November), pp. 673–91.

Majid, Abdul Rais Abdul, 2003, "Development of Liquidity Management Instruments: Challenges and Opportunities," paper presented at the International Conference on Islamic Banking: Risk Management, Regulation and Supervision, Bank-Indonesia, Jakarta, September 30–October 3.

Mauro, Paolo, 1995, "Corruption and Growth," *Quarterly Journal of Economics,* Vol. 110 (August), pp. 681–712.

Mojaverhosseini, Farshid, 2003, "An Inquiry into the Sources of Growth and Stagnation in Iranian Economy," ICER Working Paper No. 12/03 (Torino, Italy: International Centre for Economic Research).

Mullin, John, 1993, "The Equivalence of Tariffs-cum-Subsidies and Official Exchange Rate Devaluations under Dual Exchange Markets," *Journal of International Economics*, Vol. 34 (May), pp. 325–39.

Obstfeld, Maurice, and Kenneth S. Rogoff, 1996, *Foundations of International Macroeconomics* (Cambridge, Massachusetts: MIT Press).

Prasad, Eswar, Kenneth Rogoff, Shang-Jin Wei, and M. Ayhan Kose, 2003, *Effects of Financial Globalization on Developing Countries: Some Empirical Evidence*, IMF Occasional Paper No. 220 (Washington: International Monetary Fund).

Rogoff, Kenneth, 1995, "The Purchasing Power Parity Puzzle," *Journal of Economic Literature*, Vol. 34 (June), pp. 647–68.

Roubini, Nouriel, and Xavier Sala-i-Martin, 1992, "Financial Repression and Economic Growth," *Journal of Development Economics,* Vol. 39 (July), pp. 5–30.

———, 1995, "A Growth Model of Inflation, Tax Evasion, and Financial Repression," *Journal of Monetary Economics*, Vol. 35 (April), pp. 275–301.

Sarel, Michael, 1996, "Nonlinear Effects of Inflation on Economic Growth," *IMF Staff Papers*, Vol. 43 (March), pp. 199–215.

Sargent, T.J., and N. Wallace, 1981, "Some Unpleasant Monetarist Arithmetic," *Federal Reserve Bank of Minneapolis Quarterly Review*, Vol. 5 (Fall), pp. 1–17.

Schaechter, Andrea, Mark Stone, and Mark Zelmer, 2000, *Adopting Inflation Targeting: Practical Issues for Emerging Market Countries*, IMF Occasional Paper No. 202 (Washington: International Monetary Fund).

Siegel, Jeremy J., 1998, *Stocks for the Long Run* (New York: McGraw-Hill, 2nd ed.).

Sundararajan, V., and Luca Errico, 2002, "Islamic Financial Institutions and Products in the Global Financial System: Key Issues in Risk Management and Challenges Ahead," IMF Working Paper 02/192 (Washington: International Monetary Fund).

Sundararajan, V., Michel Lazare, and Sherwyn Williams, 1999, "Exchange Rate Unification, the Equilibrium Real Exchange Rate, and Choice of Exchange Rate Regime—The Case of the Islamic Republic of Iran," IMF Working Paper 99/15 (Washington: International Monetary Fund).

Sundararajan, V., David Marston, and Ghiath Shabsigh, 1998, "Monetary Operations and Government Debt Management Under Islamic Banking," IMF Working Paper 98/144 (Washington: International Monetary Fund).

Ul Haque, Nadeem, and Abbas Mirakhor, 1998, "The Design of Instruments for Government Finance in an Islamic Economy," IMF Working Paper 98/54 (Washington: International Monetary Fund).

Wakeman-Linn, John, Chonira Aturupane, Stephan Danninger, Koba Gvenetadze, Niko Hobdari, and Eric Le Borgne, 2004, *Managing Oil Wealth: The Case of Azerbaijan* (Washington: International Monetary Fund).

Walsh, Carl, 2003, *Monetary Theory and Policy* (Cambridge, Massachusetts: MIT Press, 2nd ed.).